T0225901

Snowflake Essentials

Getting Started with Big Data in the Cloud

Frank Bell
Raj Chirumamilla
Bhaskar B. Joshi
Bjorn Lindstrom
Ruchi Soni
Sameer Videkar

Snowflake Essentials: Getting Started with Big Data in the Cloud

Frank Bell
Encino, USA

Raj Chirumamilla
Monmouth Jct, USA

Bhaskar B. Joshi
Bethesda, MD, USA

Bjorn Lindstrom
Eatonville, USA

Ruchi Soni
New Delhi, Delhi, India

Sameer Videkar
Magdeburg, Sachsen-Anhalt, Germany

ISBN-13 (pbk): 978-1-4842-7315-9
https://doi.org/10.1007/978-1-4842-7316-6

ISBN-13 (electronic): 978-1-4842-7316-6

Copyright © 2022 by Frank Bell, Raj Chirumamilla, Bhaskar B. Joshi, Bjorn Lindstrom, Ruchi Soni, Sameer Videkar

This work is subject to copyright. All rights are reserved by the Publisher, whether the whole or part of the material is concerned, specifically the rights of translation, reprinting, reuse of illustrations, recitation, broadcasting, reproduction on microfilms or in any other physical way, and transmission or information storage and retrieval, electronic adaptation, computer software, or by similar or dissimilar methodology now known or hereafter developed.

Trademarked names, logos, and images may appear in this book. Rather than use a trademark symbol with every occurrence of a trademarked name, logo, or image we use the names, logos, and images only in an editorial fashion and to the benefit of the trademark owner, with no intention of infringement of the trademark.

The use in this publication of trade names, trademarks, service marks, and similar terms, even if they are not identified as such, is not to be taken as an expression of opinion as to whether or not they are subject to proprietary rights.

While the advice and information in this book are believed to be true and accurate at the date of publication, neither the authors nor the editors nor the publisher can accept any legal responsibility for any errors or omissions that may be made. The publisher makes no warranty, express or implied, with respect to the material contained herein.

Managing Director, Apress Media LLC: Welmoed Spahr
Acquisitions Editor: Jonathan Gennick
Development Editor: Laura Berendson
Coordinating Editor: Jill Balzano

Cover designed by eStudioCalamar

Cover image designed by Freepik (www.freepik.com)

Distributed to the book trade worldwide by Springer Science+Business Media LLC, 1 New York Plaza, Suite 4600, New York, NY 10004. Phone 1-800-SPRINGER, fax (201) 348-4505, e-mail orders-ny@springer-sbm.com, or visit www.springeronline.com. Apress Media, LLC is a California LLC and the sole member (owner) is Springer Science + Business Media Finance Inc (SSBM Finance Inc). SSBM Finance Inc is a **Delaware** corporation.

For information on translations, please e-mail booktranslations@springernature.com; for reprint, paperback, or audio rights, please e-mail bookpermissions@springernature.com.

Apress titles may be purchased in bulk for academic, corporate, or promotional use. eBook versions and licenses are also available for most titles. For more information, reference our Print and eBook Bulk Sales web page at http://www.apress.com/bulk-sales.

Any source code or other supplementary material referenced by the author in this book is available to readers on GitHub via the book's product page, located at www.apress.com/9781484273159. For more detailed information, please visit http://www.apress.com/source-code.

Printed on acid-free paper

Table of Contents

About the Authors

Frank Bell is a SnowPro, Snowflake Data Hero, entrepreneur, and data thought leader who has been working with databases since 1994 when he first started with Oracle in the United States Air Force. He ran the consulting firm IT Strategists from 1999 to 2019 and built one of the fastest-growing Snowflake practices in 2018, which was sold to Fairway Technologies and then acquired by Accenture. He now runs IT Strategists (ITS) Cloud Products (`https://itstrategists.com`) and Snowflake Solutions (`https://snowflakesolutions.net`), which are focused on building Snowflake tools, such as Snoptimizer (`https://snoptimizer.com`) (Snowflake Performance, Cost, and Security Optimization). He also leads the Accenture Snowflake West Market Unit. He has lived, breathed, and eaten Snowflake since early 2018 and believes Snowflake and the data cloud are one of the top game-changing technologies of this decade.

Raj Chirumamilla is a Snowflake Certified Architect and AWS Solutions Architect Associate with 20+ years of IT experience. He is an experienced, hands-on solutions architect working in the data and analytics area, helping customers migrate to cloud and build cost-effective and highly available solutions and data pipelines with modern data architecture frameworks. He has been working with Snowflake's cloud data platform (CDP) for more than three years.

Bhaskar B. Joshi is a Snowflake Certified Architect with 20 years of experience in managing data on various platforms. He has been working with Snowflake for over three years in multiple industries. He is currently a hands-on data architect helping customers create scalable solutions using modern data architecture frameworks in the cloud.

Bjorn Lindstrom is an IT veteran with 40 years of experience in many aspects of computer technologies. He has held roles in software development, software quality assurance, database administration, and large-scale big data systems and is an experienced solutions architect. He was one of the first to be certified in Snowflake.

ABOUT THE AUTHORS

Ruchi Soni is a technology leader and multi-cloud enterprise architect. Her work lies in helping customers accelerate their digital transformation journey to the cloud and build next-generation apps on forward-looking platforms such as Snowflake. She brings extensive experience in planning, architecting, building, and scaling future-ready platforms that are highly available and agile. Ruchi is SnowPro Core certified, Google Cloud certified, and AWS certified and is an Accenture Certified Master Data Architect and Senior Technology Architect.

Sameer Videkar is a data expert with more than 12 years of experience working on data platforms, enterprise data warehouse systems, master data management, data migration, and ETL implementations. He is a certified data architect with additional certifications in Snowflake, Agile Scrum, and PMP.

About the Technical Reviewers

Mike Gangler is a senior database specialist and architect. He's also an Oracle ACE with expertise in database reliability and architecture and a strong background in operations, DevOps, consulting, and data analysis. He is currently serving on the board of directors of the SouthEastern Michigan Oracle Professionals (SEMOP) and Michigan Oracle Users Summit (MOUS, www.mous.us). When not working, Mike enjoys spending time outdoors – hiking or fishing – and spending time with his family and eight grandchildren.

Chaitanya Geddam is an innovator, entrepreneur, and people person at heart. Chaitanya has worked in the Information Technology space for more than 18 years and incubated many data on cloud solutions like Snowflake. He helped enable training on Snowflake technologies for 1000+ resources, architect touchless data validation tools, and operationalize resilient migration factory solutions. Chaitanya has deep industry knowledge and business expertise in enterprise data warehouses, data lakes, and relational databases. Outside technology, Chaitanya loves to meditate, is an avid reader of spiritual books, and lives in New Jersey with his wife and two young boys.

Foreword

In 2010, Marc Andreessen wrote an op-ed in *The Wall Street Journal* titled "Why Software Is Eating the World" as many brick-and-mortar businesses were being disrupted by digitally native firms like Amazon, Uber, and Apple. The pace of disruption has accelerated further as organizations have embraced a data-driven approach to transforming businesses. Data has become central to transforming how companies interface with their customers, suppliers, and partners, find new ways to generate revenues, create new ecosystems, and improve operational efficiencies.

Data is now used in innovating almost every aspect of life, such as new drug research, autonomous vehicles, smart cities, adjusting insurance claims, creating art and music… and the list goes on. Data has become so pervasive that the best way to describe this new world is that "Data is *feeding* the world." The recent COVID pandemic has also necessitated that companies move their apps and data estates to the cloud. The combined data and cloud trend has given rise to a new breed of data platforms that enable enterprise initiatives to become data-driven. One such popular data platform is Snowflake – a platform that has taken the industry by storm since its unprecedented IPO in September 2020.

This is a timely book on the Snowflake data platform that covers the essentials as well as some expert topics such as data architectures. Frank, Bjorn, Raj Chirumamilla, Sameer Videkar, Bhaskar Joshi, and Ruchi Soni are top experts in Snowflake and have packed this book with stellar content on Snowflake and how to use it effectively.

The authors have become deep Snowflake experts and have been working extensively on Snowflake for many years. They have helped build Snowflake consulting solutions that have helped numerous Snowflake implementations and clients. Frank, in particular, has been immersed in Snowflake since the beginning of 2018. He transformed his previous big data consulting business to become one of the key Snowflake implementation partners before being acquired by Accenture.

This book will help data professionals learn all the essentials around the Snowflake Data Cloud. It also covers some of the transformative Snowflake architecture and features in depth such as data sharing. My favorite chapter in the book is Chapter 14 on data sharing, exchanges, and marketplaces. It provides an insightful look at how data sharing is transforming the evolution of data collaboration to digitally transform businesses.

—Shail Jain

CHAPTER 1

The Snowflake Data Cloud

Snowflake has been one of the most transformative data technologies I've come across in my 30-year technology career. Over the last several years, Snowflake has disrupted the big data and analytical relational database management system (RDBMS) industries. The creation of a Software as a Service (SaaS) cloud database built entirely on multiple public clouds has been an analytical database game changer. The Snowflake Data Cloud with an almost unlimited scale has been a revolutionary improvement in both data processing ease and scale.

Big Data Cloud History

As the Internet continued to grow in the decade of the 2000s, data creation and collection grew at a rapid pace. Amazon introduced S3 in March 2006 as part of Amazon Web Services (AWS). S3 was a great way to store files in the cloud, but it didn't have any traditional data management capabilities. The next month in that same year, the Apache Software Foundation introduced a new big data technology named Hadoop. Hadoop for quite some time became the go-to big data solution. Many of us who were data professionals at the time worked to implement Hadoop solutions, but initially Hadoop was very complex and required developers with coding knowledge to get any value out of it. Hadoop did not have any SQL interface until 2010. Hive was only introduced on October 1, 2010, and then it still wasn't really well integrated with the entire Hadoop solution.

Snowflake Beginnings

Realizing there were still incredibly difficult challenges scaling big data solutions in 2012, the Snowflake founders came together to build a relational database management system (RDBMS) built from the ground up on a cloud architecture.

© Frank Bell, Raj Chirumamilla, Bhaskar B. Joshi, Bjorn Lindstrom, Ruchi Soni, Sameer Videkar 2022
F. Bell et al., *Snowflake Essentials*, https://doi.org/10.1007/978-1-4842-7316-6_1

NoSQL solutions including Hadoop were cloud technologies that had come a long way since 2006, but Hadoop especially was still expensive and too complex for most organizations to operate. The Snowflake founders, Benoit, Thierry, and Marcin, were the first technologists to completely rethink and rearchitect an RDBMS to work with cloud-based technologies. This new data architecture created a major differentiation in speed and scale, ease of use, and ease of data sharing that led to Snowflake's rapid customer adoption and business growth. These fundamental technical differentiators created foundational business differentiators for customers. These business differentiations were significant enough to overcome any business and technology switching costs related to moving to the Snowflake Data Cloud.

In this chapter, we will cover what the Snowflake Data Cloud really is and why you can benefit as both a business professional and data professional from learning the Snowflake essentials. We introduce to you the overall Snowflake Data Cloud and the transformative impact it is having on the overall world of data sharing.

Why the Snowflake Data Cloud Is Different

Snowflake was the first database architected to run on the cloud from the ground up. Snowflake prefers to brand itself as the Snowflake Data Cloud since June 2020, but it was created to be a SQL database engine with automated scaling and tuning at first. This is what made Snowflake so disruptive and separated it from traditional analytical RDBMSs, on-premise massively parallel processing (MPP) solutions, and its early cloud competitors.

Google BigQuery, AWS RedShift, Microsoft Azure, and even Databricks were all created from totally different architectural foundations and with different initial purposes. RedShift was a modification of the PostgreSQL database that initially was not architected to separate compute and storage. RedShift was still the first cloud analytical database, so it still had a first mover advantage, but its foundations are built off an existing on-premise RDBMS technology. Google BigQuery was built on top of Dremel technology. BigQuery was initially designed as a black box query engine, not an RDBMS. Databricks was created as the enterprise version of Apache Spark, an open source distributed computing framework.

Snowflake differentiates itself from all of these other solutions since its core architecture was an RDBMS built for the cloud and initially created by two Oracle RDBMS veterans who took all of their advanced RDBMS engineering knowledge to create a fully scalable cloud database system. This is crucially important because the core of how a system works and grows comes from its foundational purpose and architecture (assuming it stays true to its foundational architecture).

Snowflake's Unique Architecture

Due to some of the original underlying architecture, Snowflake was able to expand beyond the concept of just a single database management system. Readers coming from RDBMSs understand the sheer pain we used to have to deal with connecting on-premise database systems to even different databases that ran on the same database systems such as Oracle and SQL Server.

Snowflake's architecture is a hybrid model of both a shared-disk and a shared-nothing architecture. At the core of Snowflake's architecture are three separate layers that we will go into more depth in Chapter 3. Here is a quick overview of them:

- Cloud Services: The cloud services layer of Snowflake handles all of the services within the database such as metadata management, authentication, security, and query optimization.

- Compute: Snowflake has virtual warehouses that run the compute. The query layer is separated from the disk storage.

- Storage: Snowflake uses micropartitions, which are heavily compressed and optimized to organize the data into a columnar data store. The data is stored within the cloud provider's cloud storage (e.g., S3 in AWS). Compute nodes connect to the storage layer to retrieve the data and process it.

Figure 1-1 shows a visualization of the Snowflake Data Cloud's three layers: cloud services, compute, and storage.

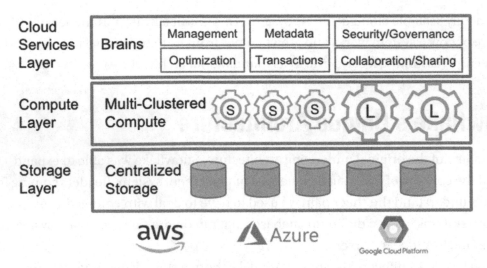

Figure 1-1. *Snowflake's Three-Layer Architecture*

Snowflake's Unique Platform Features

There are five fundamental Snowflake Data Cloud features enabling Snowflake to be so disruptive and fundamentally different from previous database solutions:

- The separation of compute from storage

- Automated data maintenance and scaling

- Ease of use

- Speed. Speed. And more speed

- Data sharing

Let's go through these fundamental differences one by one.

The Separation of Compute from Storage

The main architectural decision to separate compute from storage provided Snowflake with major differentiation from all of its competitors at the beginning besides Google BigQuery. This enabled Snowflake to come to market with a true pay-as-you-go RDBMS data service. At the time, this was simply amazing and unheard of. It was the first cloud RDBMS where you only had to pay for what you consumed. This allowed small and mid-size companies to access large compute for reasonable costs. It also enabled unprecedented scaling of compute to solve data challenges.

Any customer of Snowflake in the world could bring *massive* compute to any data problem for a few seconds or minutes at a reasonable cost vs. spending months negotiating and installing the standard RDBMS big data solutions, such as Netezza-, Teradata-, or Hadoop-based solutions. This was fundamentally revolutionary for a data engineer or data entrepreneur who was comfortable with SQL. We could be querying and analyzing large datasets within the same day with our Snowflake account. This was a game changer.

Cloud-Backed Availability

Also, this enabled a distributed architecture of availability across availability zones and removed the immediate need of all previous on-prem solutions for backup. This architecture also enabled time travel and cloning features, which were revolutionary concepts to bring to an RDBMS. (Since 2018, BigQuery and Databricks have copied the basics of Snowflake's features.)

Cloud-Enabled Scale

From an engineering perspective, what should not go unnoticed is Snowflake's micropartition architecture, which is really a continuous write structure in cloud data storage. This fundamentally creates non-locking data retrieval from storage files. This type of architecture allows for massive scale and finally allows data to only have one single source.

This is a key point of differentiation. The Snowflake Data Cloud overall brings scale unlike on-prem solutions and even other data solutions based on the cloud. Snowflake's architecture created a solution that can share data without making copies. Copies of data (including Datamarts) were a necessary architecture in the past so that data performance could scale to meet business needs. The problem is that copies of data necessitate more maintenance and create more complexity, organization, and governance challenges and costs. One of the largest business data problems across enterprises still is inconsistent data and inconsistent analysis due to analyzing different copies of data.

Snowflake Compression

Another fundamental engineering work from Snowflake was the proprietary and sizeable compression. Faster compression and less data transfer led to less Input Output (IO) costs and an overall faster architecture. It is not uncommon to see 5x compression when migrating data to Snowflake.

Automated Data Maintenance and Scaling

One of the major differentiators of Snowflake that still remains today is the fundamental change of making database maintenance, management, and scaling a business function vs. an engineering function. From 1994 to 2018, I spent too much time learning every single trick on how to optimize data architectures. It was fun to be a technical hero to come into a situation where the RDBMS was not scaling and place a few indexes and speed everything up dramatically, but it really is something that could be engineered into a database system to be automated.

Fundamentally, I assume we engineers had blinders on and overlooked that we had the performance metadata for years to create self-indexing and self-optimizing database systems. Also, for all the data professionals reading this, star and snowflake schemas and Ralph Kimball's growth in popularity came about only due to technical scale limitations. The creation of Hadoop was driven by similar limitations of big data scaling as well. RDBMSs that people loved and were comfortable using needed database administrators (DBAs) too often to maintain speed and scale even for mid-size workloads as more and more data was created, stored, and analyzed. Also, as data became bigger and bigger, traditional RDBMSs and even on-premise MPP revolutionary solutions at the time like Netezza, Teradata, and Exadata couldn't scale either. Snowflake was the first data solution to embrace internal indexing and scaling for the analytical database. This was another game changer and fundamentally changed the maintenance cost structure for organizations by removing the bulk of complexity and the people maintenance costs of DBAs and data engineers to scale basic growth and reporting RDBMSs.

Ease of Use

One of the key features to technologies that are adopted rapidly is the ease of use of the technology. Data professionals find Snowflake very easy to use if they come from any of the traditional RDBMS or MPP database backgrounds. The founders really designed Snowflake to be an easy-to-use analytical database by removing cumbersome administrative burdens, establishing all features around consistent DDL and DML SQL standard syntax, and making it very easy to join and share data.

SQL Is Well Known

By using the RDBMS common SQL language as the core data retrieval mechanism, Snowflake made it much easier for the millions of data professionals proficient with SQL to interact with and understand their offering. SQL has been around since 1974 and is used by millions of people across the world. Many more people know SQL than Python and other programming languages. This has enabled SQL data professionals to adopt the platform more easily vs. other solutions initially not based on ANSI SQL.

Joining Enterprise Data Is Easy

With on-premise databases, it was often a challenge both administratively and with performance to join data from datasets within different databases. Again, this often resulted in making copies of data and transferring it even for internal purposes. Overall, this just created more friction and work for data professionals to get things done. Snowflake removes that barrier by allowing data querying and analysis on one primary set of data. That data table never has to be copied. This is another game changer.

Viewing Past Versions Is Built In

Time travel and Zero Copy Cloning make it easier to look at your database as it existed at any point in the past of your choosing. We will go into time travel and Zero Copy Cloning in more detail later, but just realize these fundamentally changed how data professionals worked. If you made a mistake before with traditional RDBMSs, then you often would have to restore a backup and fix your mistake. With a few lines of time travel code, you can easily go back to the version of your table before you made your mistake. This enables data professionals to fix data errors within minutes vs. hours or days.

Zero Copy Cloning created the ability to truly move toward Agile Data Warehousing. For the first time with big data, data professionals were able to clone production-size data to perform full-scale production data–level tests.

Speed. Speed. More Speed Please

In the beginning of 2018, I took Snowflake to one of my long-standing clients. They had been running this massive Athena job that was taking like a month to complete. We moved a subset of that equivalent job that was taking over 24 hours over to Snowflake during a proof of concept. It finished in less than 1 hour, and we ported it over in a couple days. A more than 24x improvement without any tuning and optimizing! Thousands of Snowflake customers repeated this type of nirvana over the past several months and years. New Snowflake customers often see massive speed improvements of 2–10x++ that create massive differentiation from its data processing competition.

Data Sharing

Over the last 20+ years, data professionals used many different mechanisms to perform data sharing within organizations and across organizations. We mainly used copying techniques to copy data from one database or data warehouse to another. Constraints surrounding these techniques created entire businesses and engineering solutions around how to use architectures, tools, etc. to copy data and scale data usage and data movement.

Many data professionals still unfortunately have not been exposed to this incredibly more efficient solution of no-copy data sharing provided with Snowflake. One of the best overall features of the Snowflake Data Cloud and a Snowflake account by itself is how easily you can connect different datasets (tables, views) to any Snowflake database or schema within your account that you have access to. Data sharing is also being improved so you can easily replicate data from one account within one region and cloud provider to another region and another cloud provider.

Timeline of the Snowflake Data Cloud Creation

The story around the architecting of Snowflake goes back to 2012. The following are some of the key dates in Snowflake's development that you may want to be aware of and that might help you in talking with and selling your own clients on the platform:

- Founded - July 23, 2012

- Raised Series A Round of $5M led by Sutter Hill Ventures – August 2012

- Bob Muglia appointed as CEO – June 2014

- Raised $26M additional funds – October 2014

- Came out of stealth mode with 80 customers on AWS – October 2014

- Raised an additional $45M and launched its first product, the cloud data warehouse – April 2017

- Raised Series $100M of additional funds –

- Launched on Azure,

- raised $263M of additional funds at a 1.5B valuation - July 12, 2018

- Frank Slootman joined as CEO – May 2019

- Launched on Google Cloud Platform (GCP) – June 4, 2019

- *Launched the Snowflake Data Exchange (which eventually became the Snowflake Data Marketplace) – June 2019

- Raised another $479M in a round led by Dragoneer Investment Group – February 7, 2020

- Snowflake IPO (Initial Public Offering), which raised $3.4B, making it the largest software IPO – September 16, 2020

Summary

On September 16, 2020, Snowflake became the largest software Initial Public Offering (IPO) in history. It continues to gain additional customer adoption, and Snowflake engineering and analytics resources are scarce. If you learn these essentials provided within this book, you will be in a good position to solve Snowflake development tasks and be hired for Snowflake work.

The trend is clear that companies and entire industries are moving to the cloud. As a data professional who wants to keep up, it is essential that you continue to develop your skills with cloud databases. Snowflake currently is one of the most popular cloud databases, and you can benefit by learning how to use the essentials to make yourself more valuable for database professional jobs.

CHAPTER 2

Snowflake Quick Start

In this chapter, we will cover how you can quickly get started with Snowflake by guiding you through the account signup process. This includes choosing a Snowflake cloud provider, region, and edition. Once you have determined these details, we will provide the essentials of connecting to their web interface. We hope this chapter gets you started quickly and helps you avoid any confusion so that you are up and running on the web interface in a few minutes.

Creating a Snowflake Account

Now that you understand the fundamental differences with the Snowflake Data Cloud, let's get you started by creating your Snowflake account. Go to `https://signup.snowflake.com/` on your web browser, as shown in Figure 2-1. Snowflake currently still offers $400 of free credits over 30 days. This makes it extremely convenient for you to get started on Snowflake and learn the essentials at no or low cost. (This is a lot better than the old days of having to buy and install Oracle, Netezza, Teradata, etc. It is also way easier than installing open source data engines.) Just remember at the end of the 30 days, Snowflake will start charging you.

© Frank Bell, Raj Chirumamilla, Bhaskar B. Joshi, Bjorn Lindstrom, Ruchi Soni, Sameer Videkar 2022
F. Bell et al., *Snowflake Essentials*, https://doi.org/10.1007/978-1-4842-7316-6_2

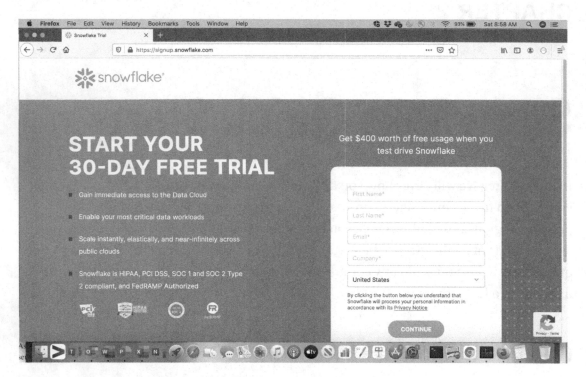

Figure 2-1. *Initial Snowflake Account Signup*

Let's get started. Fill in your details of First Name, Last Name, Email, Company, and Country and click the Continue button. The next screen (Figure 2-2) immediately prompts you to select an edition, cloud provider, and region. All of these decisions are actually related to each other, so let's cover what are the major differences between editions, clouds, and regions.

Note   Each and every account is in a specific cloud provider and region. Currently, you can only move data between different regions. Shares only can be made within a specific region. When you design your architecture, make sure to take this into account that if you want to easily share data throughout an organization, you would want to create accounts in the same region. If you do not, then you will need to replicate data to the other region, and there will be some replication cloud service costs for this.

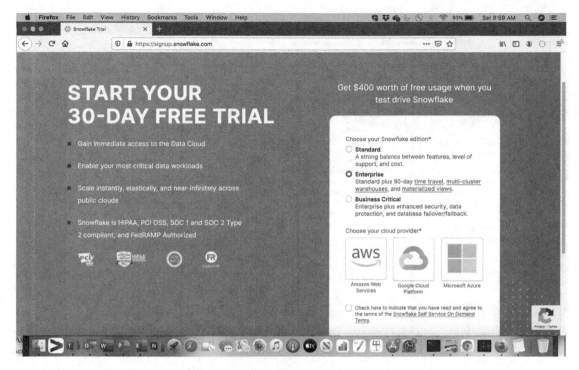

Figure 2-2. *Choosing an Edition, Cloud Provider, and Region Form*

Choosing a Snowflake Edition

The main decision is what Snowflake edition to run. The main difference is the cost related to the different pricing of credits and what features are available. Snowflake currently has four editions. Figure 2-3 shows the current available Snowflake Data Cloud editions.

13

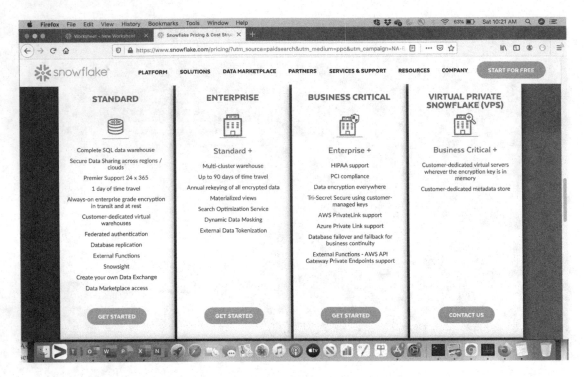

Figure 2-3. *Snowflake Editions Overview*

Snowflake Editions Overview

Snowflake comes in several editions that are priced differently and that offer different levels and combinations of features. The following are the available choices and a description of what each provides:

> Standard Edition: "A strong balance between features, level
> of support, and cost." The Standard edition is Snowflake's
> introductory offering. This offering can work well for standard
> analytical database users if they do not require features provided
> in the Enterprise edition and above. This lowest-cost Standard
> edition is actually pretty good for many businesses. I have done a
> ton of data work on Standard editions that scaled and performed
> really well. There are some key features that are not available on it
> that may push you to another edition, but it works extremely well
> for standard workloads. Any smaller organization that couldn't

afford standard on-premise expensive systems like Netezza, Teradata, and Exadata will be extremely pleased with the ease of use, scalability, and performance of this edition.

Enterprise Edition: "Standard plus 90-day time travel, multi-cluster warehouses, and materialized views." Enterprise is really Snowflake's main offering and the one that is sold to most organizations by Snowflake sales. The Enterprise edition includes all the features of the Standard edition and adds the ability to set up to 90 days of time travel, multi-cluster warehouses, and other enterprise-level cloud services such as materialized views. This allows organizations that have enterprise-level needs to have more time travel, the ability to scale out with multi-cluster virtual warehouses, and other features like search optimization services and materialized views. The main additions with the Snowflake Enterprise edition are

1. Multi-cluster virtual warehouses

2. Search optimization service

3. Materialized views

4. Database failover and fallback between Snowflake accounts for business continuity and disaster recovery

5. Extended time travel (currently up to 90 days)

6. Periodic rekeying of encrypted data

7. Column-level security

If you think you will want to use any of these features, I suggest you start with the Enterprise edition for your initial signup and trial. Just realize that this edition is typically more expensive than the Standard edition.

Business Critical: "Enterprise plus enhanced security, data protection, and database failover/fallback." We typically see customers who are larger and want enhanced security choosing this edition. The main additions with the Business Critical edition are

1. Customer-managed encryption keys (Tri-Secret Secure)

2. Support for secure and direct proxy to your other on-premise data centers or virtual networks using AWS PrivateLink or Azure Private Link

3. Support for PHI data

4. Support for PCI DSS

Even though the self-service interface only shows the three major editions above, there is one additional edition that cannot be self-service provisioned. This is the Virtual Private Snowflake (VPS) edition.

Virtual Private Snowflake (VPS) Edition: This edition is where Snowflake provides the highest level of security by providing a completely isolated and separate Snowflake environment from all other accounts. These VPS accounts do not share any resources whatsoever with accounts outside the VPS. (This edition is not offered for self-provisioning.) Organizations with highly sensitive data needs and business requirements often choose this edition. Since it uses separate resources, it is the highest-priced edition. You must realize though that while you gain extra security, this also limits the access to any other accounts and data. Also, VPS accounts use different naming conventions and have different URL structures for access. The major difference with this version is that there is a dedicated and isolated metadata store as well as pool of virtual servers for your organization's data system.

Note   In older documentation and articles, you will see reference to a "Premier" edition of Snowflake. This edition has been eliminated.

We have covered the Snowflake editions in detail here, but Snowflake will continue to make updates and changes. Please refer here for the latest information on Snowflake edition offerings: `https://docs.snowflake.com/en/user-guide/intro-editions.html#overview-of-editions`.

Selecting a Cloud Provider and Region

Choosing a Cloud Provider

Snowflake currently runs on the following three separate cloud providers:

1. Amazon Web Services (AWS)

2. Microsoft Azure

3. Google Cloud Platform

Snowflake was completely architected on the cloud, and the initial version of Snowflake only worked on AWS before June 2018. In 2018, Snowflake Corporation rebuilt the architecture to work on Azure. In June 2019, Snowflake launched on Google Cloud Platform in preview at their Snowflake Summit. For the most part, the Snowflake interface and performance to the end user feels exactly the same. When I first went onto my beta Azure Snowflake instance, I couldn't really tell the difference from the front end. Behind the scenes though, AWS is by far the most mature and widely used since Snowflake engineers built the initial version on that starting in 2012. Also, if you require your account to have HITRUST CSF certification, this is only available on AWS and Azure regions at this time of publication. See here for the latest:

`https://docs.snowflake.com/en/user-guide/intro-cloud-platforms.html#hitrust-csf-certification`

(Snowflake on Google Cloud Platform does not support HITRUST CSF.)

You may already have been using one or more of these three cloud providers for your business needs and have your minds made up. From a high-level perspective, Snowflake works almost the same from the front end on any of the clouds except for a few differences noted in the following. If you are new to the cloud, then there are a few differences in what Snowflake supports with different providers.

There are certain limitations on Azure Private Link on Azure Cloud. If this is an important feature for you, then please refer to the latest information at `https://docs.snowflake.com/en/user-guide/privatelink-azure.html#label-pl-azure-reqs-limits`.

Also, not all third-party tools in Partner Connect support Azure Cloud.

GCP does not offer any equivalent service such as AWS PrivateLink or Azure Private Link for configuring a direct secure connection between virtual private clouds. Snowflake on GCP is not currently certified for HITRUST CSF.

Choosing a Region

You picked what cloud provider you ideally want to run on. Now it is time to pick a region. In Figure 2-4 are the current regions. Most users choose a region based on their locality and their business needs. Currently, if you have a requirement to have your Snowflake Data Cloud hosted in Asia, then you must choose between AWS and Azure since GCP is not supported in any Asia regions currently. What providers and regions are offered is often being updated and changing. You can check here for the latest regions offered:

`https://docs.snowflake.com/en/user-guide/intro-regions.html`

Currently, Snowflake has these regions supported on these three specific cloud providers as displayed in Figure 2-4.

AWS	Azure	GCP
US West (Oregon)	West US 2 (Washington)	US Central1 (Iowa)
US East (Ohio)	Central US (Iowa)	Europe West2 (London)
US East (N. Virginia)	East US 2 (Virginia)	Europe West4 (Netherlands)
US East (Comm. Gov. N. Virginia)	US Gov Virginia	
Canada (Central)	Canada Central (Toronto)	
EU (Ireland)	North Europe (Ireland)	
Europe (London)	West Europe (Netherlands)	
EU (Frankfurt)	Switzerland North (Zurich)	
Asia Pacific (Tokyo)	Southeast Asia (Singapore)	
Asia Pacific (Seoul)	Australia East (New South Wales)	
Asia Pacific (Mumbai)		
Asia Pacific (Singapore)		
Asia Pacific (Sydney)		

Figure 2-4. *Snowflake Supported Regions in Cloud Providers*

Understanding Snowflake Edition Pricing

As of this publication date, Snowflake's pricing starts at the following price points per credit in USD.

- Standard - $2/Credit

- Enterprise - $3/Credit

- Business Critical - $4/Credit

Pricing is dependent upon specific regions. For the latest pricing, please refer to Snowflake's pricing page. You just need to provide your choice of edition, cloud provider, and region to get their latest prices (`www.snowflake.com/pricing/`).

Snowflake currently offers two ways to buy their Data Software as a Service (DSaaS): on demand or pre-paid capacity. Snowflake sales is focused on selling pre-paid capacity plans of certain lengths and offers negotiated pricing depending on the size of purchase and length of agreement.

Immediately Connecting to Snowflake

Once you select your Snowflake edition, cloud provider, and region and click "Get Started," you will be taken to a screen that looks like Figure 2-5.

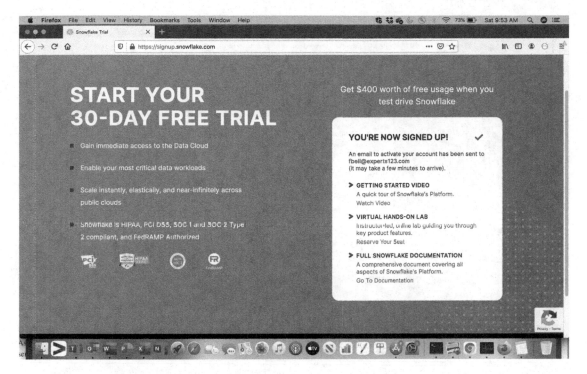

Figure 2-5. *Snowflake Account Signup Confirmation Screen*

At the same time, you will get an email titled "Activate Your Snowflake Account." This email has a standard request to confirm your email and activate your account. Once you click Activate, you will be taken to a screen where you can create your initial username and password, which looks similar to Figure 2-6.

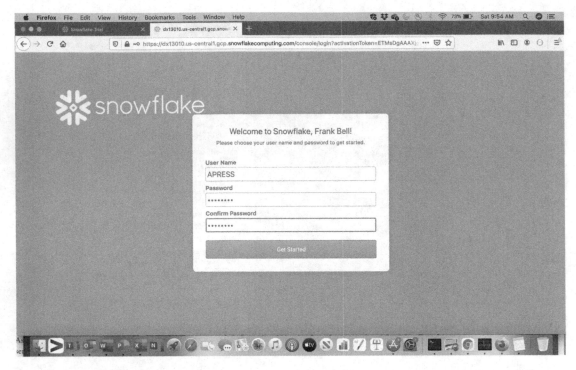

Figure 2-6. *Snowflake Initial Setup Screen*

Snowflake works on the most popular browsers of Chrome, Safari, Firefox, Opera, and Edge. Here are the specific version requirements if you start to have any problems:

`https://docs.snowflake.com/en/user-guide/setup.html#browser-requirements`

Initial Web View

If everything worked out correctly, then within a minute or so you can be up and running on Snowflake! (If you have worked with installing databases or setting up on-premise ones in the past, you will be amazed at how quickly you can be up and running and querying data.) Figure 2-7 shows what Snowflake's initial web interface will look like.

Snowflake has done a reasonably good job with its "Getting to know Snowflake" material. If you know SQL, then this is a really great way to take Snowflake for a spin by using one of their "Welcome to Snowflake" tutorials, such as "Run a query in the worksheet." If you want to get going quickly from here with the web interface, then jump to Chapter 4 where we cover in depth the Snowflake visual web interface as it works now.

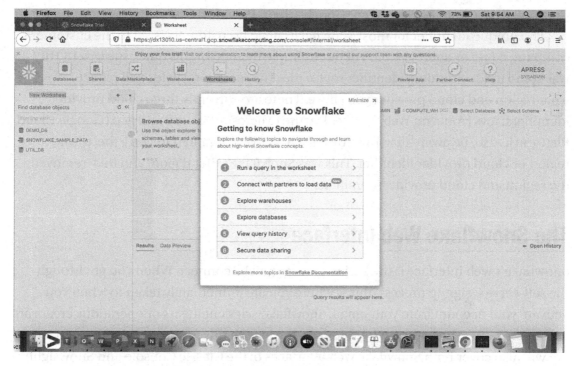

Figure 2-7. *Snowflake First Screen After Login*

Tip   When you do self-service signup and provisioning, a random URL is created for your account, and it is dependent upon the cloud provider and region you selected. We recommend immediately bookmarking this in your browser so you don't forget it. If you want to change the name to something customized and related to your company, you need to contact Snowflake operations.

Initial Snowflake Account URL

In Figure 2-7 the initial Snowflake account URL sent for this example is

    https://dx13010.us-central1.gcp.snowflakecomputing.com

Each part of the URL structure corresponds with the initial decisions you made filling out the preceding account signup forms. Let's break down the URL structure:

https://dx13010.us-central1.gcp.snowflakecomputing.com

> Account Regional Unique identifier: dx13010
> Snowflake Account Region: us-central1

Snowflake Cloud Provider: gcp (Google Cloud Provider)

The unique identifier for most Snowflake capacity customers is set to the customer's company name or abbreviation. For our example though for this quick start, dx13010 was some unique random identifier set for me. The remaining fields will reflect what region and cloud provider you chose. There are currently three cloud provider identifiers: aws (Amazon Web Services), azure (Microsoft Azure), and gcp (Google Cloud Platform). Also, be aware that for Snowflake AWS West Region (Oregon), there is no region or cloud provider identifier. This was Snowflake's first region, and they removed the region and cloud provider identifiers.

The Snowflake Web Interface

Snowflake's web interface is the easiest way to initially connect. When you go through the self-service signup process, this is where you are immediately taken to when you activate your account from your email. Snowflake sales engineers or operations create an account for you. Then you can immediately log in with this full-featured web interface. We will also cover both Snowflake web interfaces of the Classic Console and Snowsight in great depth in Chapters 4 and 5 so you know all the Snowflake interface essentials.

Summary

In this chapter, we provided details on how you can get started quickly with Snowflake. We hope it enables you to try out Snowflake rapidly by creating a trial account and getting started rapidly within minutes by making it easy to select a cloud provider, edition, and region. Once your account is created, you can jump into the web interface and begin creating objects and loading data. You can jump to Chapter 4 or 5 to go through the web interface next or jump to Chapter 12 to see how to easily start loading sample datasets.

CHAPTER 3

Snowflake Data Cloud Architecture

This chapter will cover the essentials of the Snowflake Data Cloud architecture that has made Snowflake widely popular. This hybrid architecture provides Snowflake with ease of use as well as fast and scalable performance. When the founders decided to build a new relational database completely based on the cloud, they were able to create architectural advantages beyond existing database architectures. One of the key architectural beliefs they were founded on was that tying storage to compute created challenges with scaling on the cloud.

In this chapter, we will cover how Snowflake's decision to have a hybrid architecture of traditional shared-disk and shared-nothing architectures has helped Snowflake create a powerful and highly scalable RDBMS solution. Snowflake capitalizes on using a central data repository similar to a shared-disk architecture for persisted data within each cloud provider. At the same time, it processes queries using MPP (massively parallel processing) similar to shared-nothing architectures. Snowflake uses compute clusters to do this where each node in the cluster stores a part of the dataset locally. This hybrid approach provides both data management simplicity and improved performance of the scale-out architecture.

The Snowflake Data Platform as a Cloud Service

Snowflake has introduced us to DSaaS (Data Software as a Service), which runs on their data platform as a cloud service. This simply means there is no software or hardware to install, configure, or manage. There are no software upgrades to manage either. Snowflake Corporation manages all of that complexity for you.

© Frank Bell, Raj Chirumamilla, Bhaskar B. Joshi, Bjorn Lindstrom, Ruchi Soni, Sameer Videkar 2022
F. Bell et al., *Snowflake Essentials*, https://doi.org/10.1007/978-1-4842-7316-6_3

Snowflake takes care of the ongoing maintenance of your database and the tuning, querying, security, and management services related to it. This also means that you, the data professional, now have access to a full RDBMS delivered in the cloud that is optimized for scale. This really changed the landscape for organizations to not have to invest in continuous maintenance of hardware and software. It frees the data professional from having to deal with a lot of scaling engineering that was required in the past with almost all other data systems. Snowflake's unique architecture also provides a true cost and speed advantage for organizations by removing lots of underlying database administrator maintenance costs and database planning costs. Since Snowflake runs completely on the cloud, every component of Snowflake's DSaaS runs on cloud infrastructure within each cloud provider (besides the optional connectors, drivers, and command clients). In this chapter, we will cover the three main architectural layers of Snowflake's Data Software as a Service.

Caution   While the removal of most DBA activities is truly amazing and game changing (Hooray! No more index management and maintenance!), all enterprise-level database organizations and professionals who are processing terabytes to petabytes of data will quickly realize that with great data power still comes great responsibility. The ease of use that the Snowflake Data Cloud provides will still require either automated or some administrative human management of access, data cloning, data usage, data quality, and data governance. I mean it is awesome that you can load terabytes of data into Snowflake and process it quickly. In order to maintain high-quality data, an organization must use professional third-party services or customized Snowflake functions or tools to do resource monitoring and data governance. Snowflake data warehouses, lakes, and clouds can become out-of-control data swamps and cost-control nightmares if you do not set up and continually monitor your Snowflake account. [You have received fair warning here!!!]

Big Data Architecture History

Before Snowflake, the main two big data architectural approaches were shared nothing and shared disk. Figure 3-1 shows a visualization of the two different architectures. Shared nothing is when the data is partitioned and processed across separate server nodes. Each node has the sole responsibility for the data it has. The data is completely segregated. Shared disk is basically the opposite where all data is available to all the nodes. Any of the nodes can write to or read from any part of the data it wants.

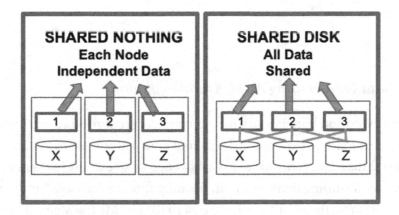

Figure 3-1. *Shared Nothing vs. Shared Disk*

Snowflake's Hybrid Architecture

Snowflake's architecture is a hybrid of both the shared-nothing and shared-disk architectures. It is set up to take advantage of benefits of both concepts. The key is that Snowflake separates storage and compute, which gives it great flexibility in both scaling processing up and out and allowing for consumption-based pricing.

The Data Warehouse Evolution

Data warehouse technology and analytical databases have been evolving over the last 30 years from RDBMSs to MPP on-premise systems. Then Hadoop and cloud data analytical systems evolved over the last ten years. This eventually evolved to Snowflake coming out with its groundbreaking architecture separating compute from storage initially on the AWS cloud provider. Figure 3-2 illustrates this evolution.

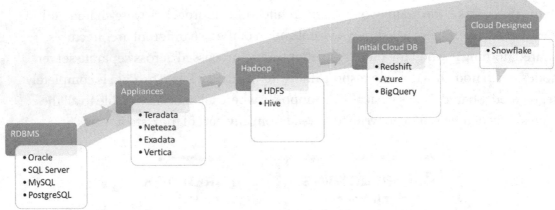

Figure 3-2. *Data Warehousing Architectural Evolution*

We saw this evolution happening with our consulting solutions over the last 20+ years, and I wrote an article explaining the overall data warehouse evolution in more depth here: www.linkedin.com/pulse/data-warehousing-evolution-frank-bell.

The Snowflake founding team saw both the migration to the cloud and the challenges related to all the existing solutions of RDBMSs, MPP systems, Hadoop, NoSQL, and initial cloud databases where the initial architecture was not from a cloud provider architectural foundation. The Snowflake founders published this white paper, which covers the fundamentals of Snowflake's architecture beliefs in the early days:

https://dl.acm.org/doi/10.1145/2882903.2903741

Figure 3-3 below shows an overview of Snowflake's three layered architecture.

Snowflake's Layered Architecture

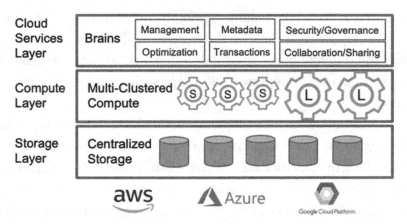

Figure 3-3. *Snowflake's Three Layers of Cloud Services, Compute, and Centralized Storage*

Let's dive into how each of these three independent layers works.

Cloud Services Layer

The cloud services layer is really the brains behind the Snowflake Data Cloud. It provides the main services of the Snowflake Data Cloud that all users interface with including

- Optimization services

- Management services

- Transaction services

- Security and governance services

- Metadata services

- Data sharing and collaboration services

    This layer controls all the authentication and security to create centralized security and better data governance. One of Snowflake's key features is that it transparently exposes the query history, and this is done through this layer as well. The services layer also handles all of the metadata management, query optimization, and Snowflake's data sharing services (which we will discuss in depth in Chapter 14).

Compute Layer

Snowflake's compute layer is fully separated from the other two layers of cloud services and storage. This compute layer runs "virtual warehouses," which can be of various T-shirt sizes and run different query workloads independently and concurrently. At a high level, these virtual warehouses are MPP compute clusters with multiple nodes of CPU and memory.

This allows organizations to have different workloads on different Snowflake warehouses. For example, one warehouse can be focused on loading data, while another warehouse is handling queries for data analysts. A Snowflake customer can technically run tens or hundreds of separate independent warehouses running different workloads and never contending for the same compute resources. These workloads can even be accessing the same exact data at the same time with no contention or bottlenecks of traditional databases. All the provisioning of this virtual compute node is done by Snowflake depending on the selections of the end user around virtual warehouse size and cluster size between one and ten nodes. All these virtual warehouses (and there can be hundreds or thousands of them if your business needs it) work completely independently.

Storage Layer

Snowflake's storage layer is completely separated from the compute layer, which also allows Snowflake to charge a more reasonable cost for storage than previously seen in the database offerings when both compute and storage were tied to each other. Snowflake's storage layer cleverly leverages the native blob storage capabilities of a cloud provider. (On AWS it uses S3. On Azure it uses Azure Blob. On Google it uses Google Cloud Storage.) This is done through a columnar database architecture, which has raw files compressed and encrypted within the native cloud storage available. This layer is designed to provide sub-second query response times with centralized data at petabyte scale. Snowflake also leverages their raw micro-partition storage technology which we will discuss in more depth later. Heavy compression on the storage layer is also used to improve performance. We often see three to five times compression with data loaded into Snowflake.

The Separation of Storage from Compute

The major architectural advantage Snowflake harnessed from the cloud was the separation of compute from storage. This is the bottleneck that every on-premise system would run into as data continued to grow bigger and bigger. Even with optimized hardware, on-premise systems at some point just could not keep up with the massive scale of cloud-based provider server farms.

This separated architecture also enabled Snowflake to deliver to customers a "pay for what you use" offering. This traditionally has been a business architecture winning formula because now for the first time even small startups could afford this pay-as-you-go architecture as they worked within investment funding to achieve product market fit and scale their data and business.

Micropartitions and Their Use in Snowflake

Micropartitions are another one of Snowflake's key architectural concepts designed to work well in a cloud architecture. The main benefits from using them are the speeds at which most workloads can be delivered compared with on-premise or other cloud RDBM systems that used traditional indexes or hardware optimizations. Snowflake automatically divides and groups rows of tables into these compressed micropartitions of 50–500 MB of data. Figure 3-4 shows an example of three micropartitions. Micropartitions are immutable physical files that are automatically partitioned based on ingestion order unless you set up auto-clustering to define how partitions should be set up. Ideally the micropartitions are clustered (sorted) as efficiently as possible to allow for **pruning**. (See Figure 3-5.) These micropartitions allow the Snowflake engine to easily replicate segments evenly for distribution across nodes. These micropartitions also are part of the architectural design that allows Snowflake to handle datasets of any size (even petabytes) since they cleverly distribute them into these micropartitions with metadata automatically and continuously updated on them.

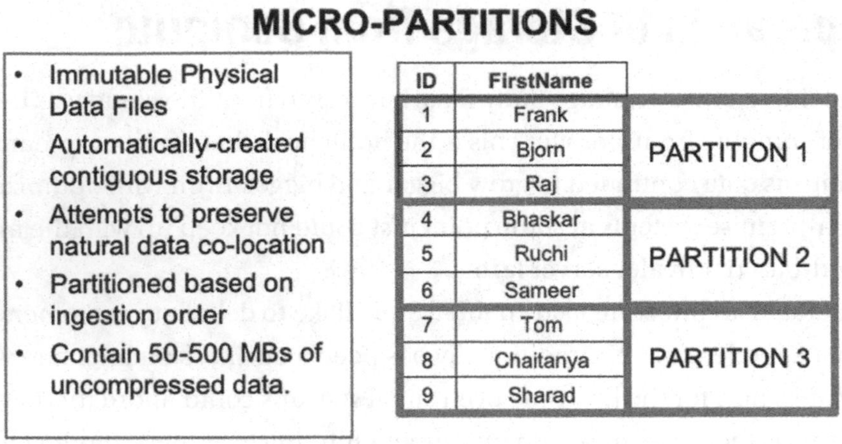

MICRO-PARTITIONS

- Immutable Physical Data Files
- Automatically-created contiguous storage
- Attempts to preserve natural data co-location
- Partitioned based on ingestion order
- Contain 50-500 MBs of uncompressed data.

ID	FirstName	
1	Frank	
2	Bjorn	PARTITION 1
3	Raj	
4	Bhaskar	
5	Ruchi	PARTITION 2
6	Sameer	
7	Tom	
8	Chaitanya	PARTITION 3
9	Sharad	

Figure 3-4. *Snowflake's Micropartitioning Example*

How Pruning Works

Query pruning is mainly what it sounds like. A database architecture is constructed to use a query optimizer that prunes away micropartitions unnecessary to run a query. This optimizes the Input Output (IO) overhead, compute, and overall work required and makes queries much faster if they only need to access a small subset of partitions vs. scanning them all. Snowflake's metadata is continuously updated and enables Snowflake's query optimizer to precisely prune columns at query runtime. This is really neat since it enables just-in-time pruning and optimization based on the micropartition metadata, which is continuously updated. Snowflake also architected this to work on semi-structured data like JSON and XML.

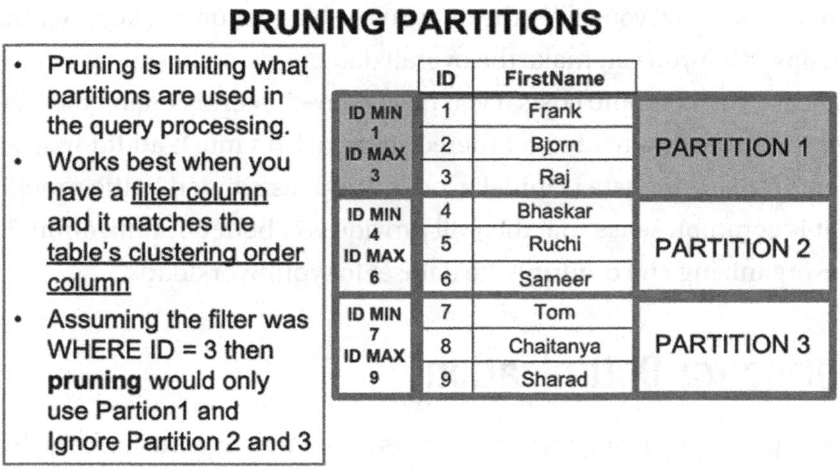

PRUNING PARTITIONS

- Pruning is limiting what partitions are used in the query processing.
- Works best when you have a <u>filter column</u> and it matches the <u>table's clustering order column</u>
- Assuming the filter was WHERE ID = 3 then **pruning** would only use Partion1 and Ignore Partition 2 and 3

	ID	FirstName	
ID MIN 1	1	Frank	
ID MAX 3	2	Bjorn	PARTITION 1
	3	Raj	
ID MIN 4	4	Bhaskar	
ID MAX 6	5	Ruchi	PARTITION 2
	6	Sameer	
ID MIN 7	7	Tom	
ID MAX 9	8	Chaitanya	PARTITION 3
	9	Sharad	

Figure 3-5. *Snowflake's Partition Pruning Example*

Cluster Keys

Cluster Keys and Automated Clustering in Snowflake

All Snowflake editions automatically cluster your data with default cluster keys when the data is ingested into tables. Typically, this is done on columns of temporal data types such as date and timestamp since this is a natural load sequence for any time series–type datasets. The reality is though that not all workloads are time-based ordered. Some tables are sequenced on some type of primary ID or joint set of columns, which organizes the sequence of the dataset within a table. Snowflake suggests that when you have tables larger than 1 TB, you need to define optimized cluster keys and enable auto-clustering. This will help if your table continues to have ingestion organization different than your workloads or Data Manipulation Language operations (UPDATE, DELETE, MERGE, etc.) that reorganize your data in non-optimal micropartitions. Just realize that while automated reclustering has many benefits including ease of maintenance and non-blocking organization, it comes at a credit consumption cost.

Tip   When you define multi-column clustering keys for a table with the CLUSTER BY clause, then the best practice is to order columns from lowest to highest cardinality.

When you load data, if you order the data before loading on the keys or filters that you will be using, then you can make the overall database system run more efficiently. This will also save you compute costs if you have auto-clustering enabled on the cluster keys because the rows are already preordered so there isn't much additional auto-clustering required. If your data is initially loaded and distributed in the order it will be queried, it is common sense that this will provide you better optimization. You are basically pre-organizing and ordering the dataset for your workloads.

Reclustering for Optimization

Reclustering is just reorganizing micropartitions based on your cluster keys. In a way it's like re-indexing or reorganizing files so that the metadata and the partitions themselves are highly optimized for pruning based on your historical workloads. Clustering and reclustering in Snowflake is now fully automatic. (If you see references to manual clustering, that has been disabled.)

Snowflake's Caching Architecture

One of my favorite features when I was initially introduced to Snowflake was that they would cache query results for 24 hours and not charge customers for accessing those query results. When you or another user **in your account** initiates the same exact query the second time, it returns instantaneously for no additional cost. I really thought this was a customer-first type of offering to not charge the end customer additional costs if Snowflake themselves did not incur costs. If you have hundreds of users doing the same exact query, you are saving tons of extra duplicative workloads that have both a hard cost and energy/climate cost.

As we discussed previously, Snowflake operates three independent redundant layers. The centralized storage layer is the cold base storage. This layer is optimized with the micropartition architecture and pruning discussed previously. Let's cover how the layers of caching and storage work together to achieve optimal performance.

The following are in Snowflake storage and caching layers:

> Result Cache: Snowflake caches the results of every query
> executed within the last 24 rolling hours. This cache is available to
> any other user on the same account who executes the same exact
> query if the underlying data has not changed.

Local Disk Cache: The virtual warehouse compute layer optimizes a separate cache as well when it is activated to retrieve and compute data operations. For example, on Snowflake AWS, each of the EC2 instances has RAM and an SSD disk. When a user runs a query, the data is retrieved from the centralized cold storage (S3) into the EC2 instance in both memory and SSD. Since Snowflake uses columnar and smart micropartitions, it will typically retrieve a limited number of columns that will be cached in the SSD disk. It is a smart limited cache based on workload patterns. This creates a warm cache that executes many regular predicted workloads extremely fast since retrieval from memory and SSD is much faster than the centralized blob storage.

Remote disk (S3, Azure Blob, Google Cloud Storage): This is where the raw compressed Snowflake micropartition files are stored.

The Benefits of Cloning

One of my most favorite features of Snowflake is the capability to clone databases within seconds. By introducing this feature, Snowflake finally allowed data engineering professionals to do truly Agile Data Warehousing. Before this feature, Agile Data Warehousing really was a misnomer when dealing with big data. Even the largest Fortune 100 companies typically did not want to pay to create a duplicate test and staging environment with on-premise or other cloud databases that required copying data. Also, when copying tens or hundreds of terabytes of data, it always is as fast as the amount and IO (Input Output) you have available, so it is less agile.

Cloning really enabled the data warehouse and big data professionals to replicate what had been working in agile software development for years. We discuss Snowflake cloning in more depth in Chapter 9.

Tip   The actual time to clone a database is dependent on the amount of metadata objects it has. Most small to mid-size databases with a few hundred objects can be cloned within a minute. Databases with large numbers of objects in the thousands will take minutes to clone.

Performance Optimization Features

As of this writing, there are three main standard optimization features in production within Snowflake. We will briefly touch on them here, and they will be covered in more depth in later chapters. While Snowflake is much more powerful than previous RDBMSs, it still can be optimized for performance. Currently, Snowflake does optimize queries with a query optimizer. Snowflake's standard performance on most data structures smaller than 1 TB even without clustering or ordering is still amazingly fast. When tables get larger than 1 TB though, the distribution of cluster keys will improve performance. Also, there is an additional query optimization service, which is in private preview currently named the Query Acceleration Service. This service will allow for larger-scale scan-heavy workloads to be accelerated on specific warehouses where this is enabled. It will be best for speeding up performance with short needs for larger-scale queries so it doesn't impact the other workloads. It also will be able to accelerate long-running queries. Let's dig into the current performance optimization services available in production.

Materialized Views

Materialized views are somewhat standard in RDBMS architectures. They allow you to store frequently used aggregations and results to avoid recomputing and increase the speed of results. Snowflake's materialized views are actually limited compared with some other vendor offerings because they *only* support materializing a subsection of an existing table. They do not provide functionality for joining tables and materializing the results. Snowflake's materialized views also have additional costs.

Automated Clustering on Cluster Keys

Snowflake automatically reorganizes tables to work with query patterns. This allows your database to use pruning more effectively to process only relevant partitions from large tables. The end benefit is faster queries and lower compute costs for the queries. There is a cost for the background clustering maintenance to re-cluster new data as it comes into the table.

Search Optimization Service

Snowflake provides this service to enhance performance around pinpoint lookups of filtered values. This is a serverless function managed by Snowflake, which creates equality predicates. This service supports data types of numbers, date, time, strings (exact match), and binary (exact match).

Summary

The Snowflake Data Cloud consists of three separate layers of cloud services, compute, and storage. Snowflake's usage of caching layers enables some of Snowflake's powerfully fast performance. Snowflake's architecture also enables groundbreaking features such as time travel and cloning, which are enabled by the micropartitions and the separation of compute from storage. The Snowflake architecture continues to add additional cloud services such as materialized views, automated clustering, and search optimization. We hope you enjoyed this chapter explaining the essentials of the Snowflake Data Cloud architecture.

CHAPTER 4

Snowflake Web Interface: Classic Console

In this chapter, we will cover Snowflake's Classic Console web interface and all the functionality within it. We will guide you to where all the key features are on the web interface. Snowflake's Classic Console is a well-thought-out web interface that has been a key part of the Snowflake platform since the beginning. As of this writing, the Classic Console is still the main interface used to interact with Snowflake. However, Snowflake Corporation has recently re-released their new front-end interface named the Preview App (Snowsight), which will be the future interface eventually. We also cover the Preview App (Snowsight) in the next chapter.

Web Interface: Classic Console Main Overview

The Snowflake Classic Console is the main way you can interact with the Snowflake Data Cloud. It provides access to almost all the functionality of the entire Snowflake Data Cloud service at both an account and an organization level. You can get access to the main areas of Databases, Shares, Data Marketplace, Warehouses, Worksheets, History, and Account (with the ACCOUNTADMIN role or equivalent). Before we cover the Classic Console, one major tip that you must always be aware of is that the interface changes are dependent upon what role you have selected in the upper right. In Figure 4-1 is the Snowflake Classic Console interface with the upper-right role selection displayed.

© Frank Bell, Raj Chirumamilla, Bhaskar B. Joshi, Bjorn Lindstrom, Ruchi Soni, Sameer Videkar 2022
F. Bell et al., *Snowflake Essentials*, https://doi.org/10.1007/978-1-4842-7316-6_4

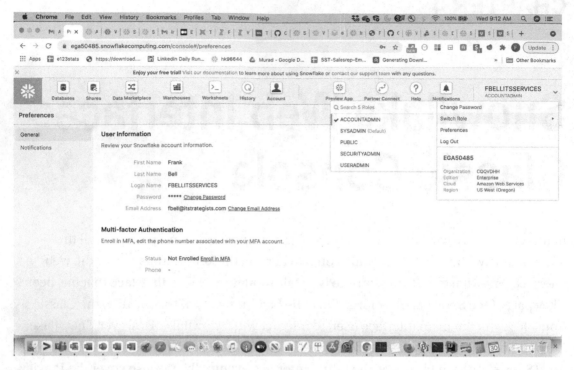

Figure 4-1. *Classic Console and Role Selection*

Tip The "role" you select in the upper right of the preceding web interface impacts what icons are displayed on your web interface. Unless you are using the ACCOUNTADMIN role (or an equivalent custom role with Account privileges), then you will not see the Account icon, and you will not be able to access any of the Account details in the Classic Console. You also will not see the Notifications icon or be able to create shares or view them without additional access.

Databases

The Databases icon is the main area for all database objects in the Snowflake Data Cloud. The main database interface contains the functionality to Create, Clone, Drop, and Transfer Ownership on databases within the Snowflake Data Cloud.

Figure 4-2 shows what the Snowflake Classic Console database listing interface looks like. You can easily create, clone, drop, or transfer ownership on databases from the buttons in the following interface. If you are coming from other relational databases, you will notice how easy it is to create a database without dealing with any complex configuration settings.

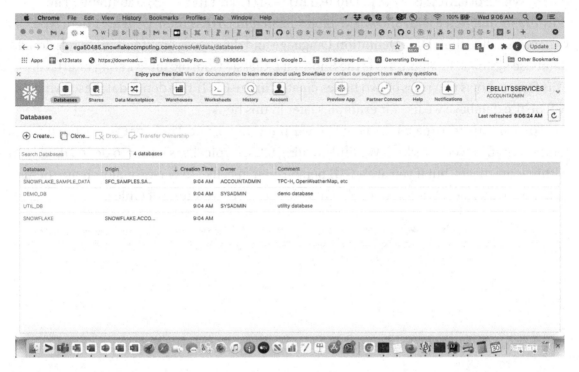

Figure 4-2. *Databases*

You may notice at first in the interface that the Create and Clone links are active and the Drop and Transfer Ownership links are grayed out. The Snowflake Classic Console is consistent in graying out actions that are not available unless you highlight the row of the listed object.

For you to view all the navigation to objects within a database such as tables, views, schemas, etc. in the Classic Console, you need to click a database name on the left-hand side of the database listings. Once you click the database name, you will see the navigation change to display tabs for the main seven objects within databases: Tables, Views, Schemas, Stages, File Formats, Sequences, and Pipes. You will also notice the navigation displays which database you have selected with the Databases ➤ [Database Name] in the interface. Let's go through each of these database objects and how to navigate to them and when you want to use them in your Snowflake Data Cloud work.

Tables

Tables are the key construct of all relational database systems. They are the mechanism that contains the data. Snowflake tables are easy to use if you come from any previous relational database work since they are created by normal Data Definition Language (DDL) syntax of CREATE, ALTER, DROP, TRUNCATE, and RENAME statements. The only major difference with Snowflake tables related to DDL is their VARIANT data type. Otherwise, all of their Data Definition Language and data types are standard to relational database systems. In Figure 4-3 is the Snowflake Classic Console table listing interface. This figure shows two rows of two tables created in the CITIBIKE demo database, which will be the database we use for examples later in this book.

You can Create, Create Like, Clone, Load Table, Drop, and Transfer Ownership on tables from this web interface. We find creating tables typically is easier to do in DDL within worksheets, but if you are a GUI person, you may like to occasionally create them in the web interface, but this doesn't really scale to large amounts of tables.

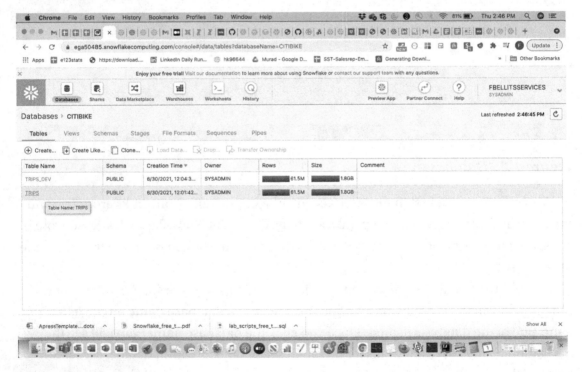

Figure 4-3. *Tables*

You will also notice again that some of these functions are grayed out. Once you click within the row of the table, then they will become active like how the database object listings worked.

Views

Views are another key construct of relational database systems, allowing users to create virtual and materialized views of physical table data. The main purpose of views in relational database systems is to provide additional security of data and allow for flexibility of view changes without having to move/copy data. Snowflake also provides a feature of secure views, which allow users to create views with the definition and the details of the view hidden from the end viewer/user for greater security. Secure views also prevent exposing underlying data to user-defined functions or other programmatic mechanisms. Secure views are the ONLY type of view you can use to share in Snowflake's data sharing feature. Secure views should not be used for views that do not need this level of privacy or security because Snowflake's query optimizer bypasses some optimizations used for regular views and there could be some level of query performance impacts with secure views.

Snowflake also provides their version of materialized views for Enterprise and above editions, which can help provide fast performance of queries by materializing the data within the view. Snowflake's version of materialized views can only be created from one table. You cannot join other tables to create a Snowflake materialized view.

A list of views is shown in Figure 4-4 where you can take actions such as Create, Drop, and Transfer Ownership of views. Like the table listing, you will notice that certain functions are grayed out that need to be done on one specific view. You must select a view in the list to have the Drop and Transfer Ownership links become active.

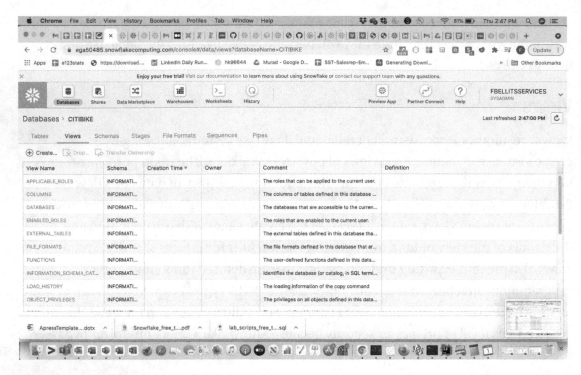

Figure 4-4. *Views*

Schemas

Schemas are a relational database mechanism created for organization and security. They are a common mechanism used within relational databases and work similarly to any relational database you have used in the past. Snowflake schemas are also enhanced with the key new Snowflake feature of cloning. Figure 4-5 shows a standard view of Snowflake schemas listed within a database.

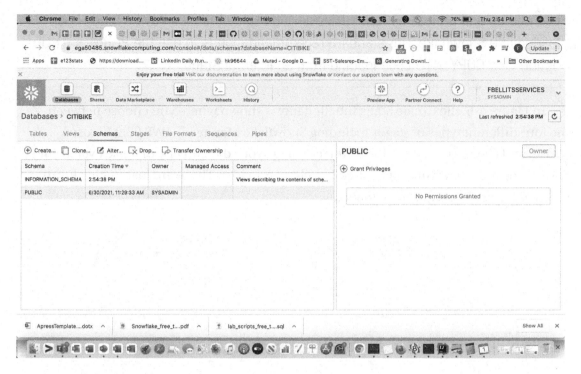

Figure 4-5. Schemas

Stages

Stages are a Snowflake Data Cloud concept and specific to Snowflake. All cloud databases require you to "stage" or really move the data from on-prem or other locations to an accessible cloud location. Snowflake's unique architecture allows you to have stages in four different ways, which include an "Internal" Snowflake Stage (technically on the AWS cloud but completely Snowflake controlled) and "External" Stages, which currently include Amazon S3, Microsoft Azure, and Google Cloud Platform.

Tip   One key point to be very clear about is that when you transfer files to an Internal Snowflake Stage, you are charged for that file storage as a part of Snowflake's storage costs. We have seen many Snowflake users stage large amounts of files in Snowflake Stages and load them into the database and forget to purge them from the Internal Stage. If these files are sizeable, the storage costs can add up and provide no real business value since the files have already been loaded into Snowflake. Our standard recommendation is to use the PURGE = TRUE option with COPY INTO code loading from an Internal Stage.

In Figure 4-6, the Create Stage functionality is shown where you choose between the four different types of stages including Snowflake Managed, AWS S3, Microsoft Azure, and Google Cloud Platform. You can stage your data on any of these even if your Snowflake account is running off another cloud provider. These are just staging locations where data can be staged to be loaded.

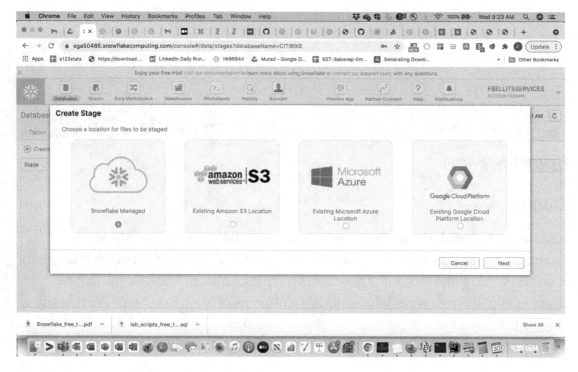

Figure 4-6. *Stages*

Let's understand how you create an External Stage on Amazon's AWS S3 file storage. Every External Stage needs a stage name and stage URL at a minimum. Most customers want to have encryption security as well for almost all use cases, so you can easily add encryption keys.

Figure 4-7 shows the interface you can use to add all of these elements of the AWS External Stage depending on your business and technical needs including Name, Schema Name, S3 URL, AWS Key ID, AWS Secret Key, Encryption Master Key, and Comment. I always recommend that you comment every object you create.

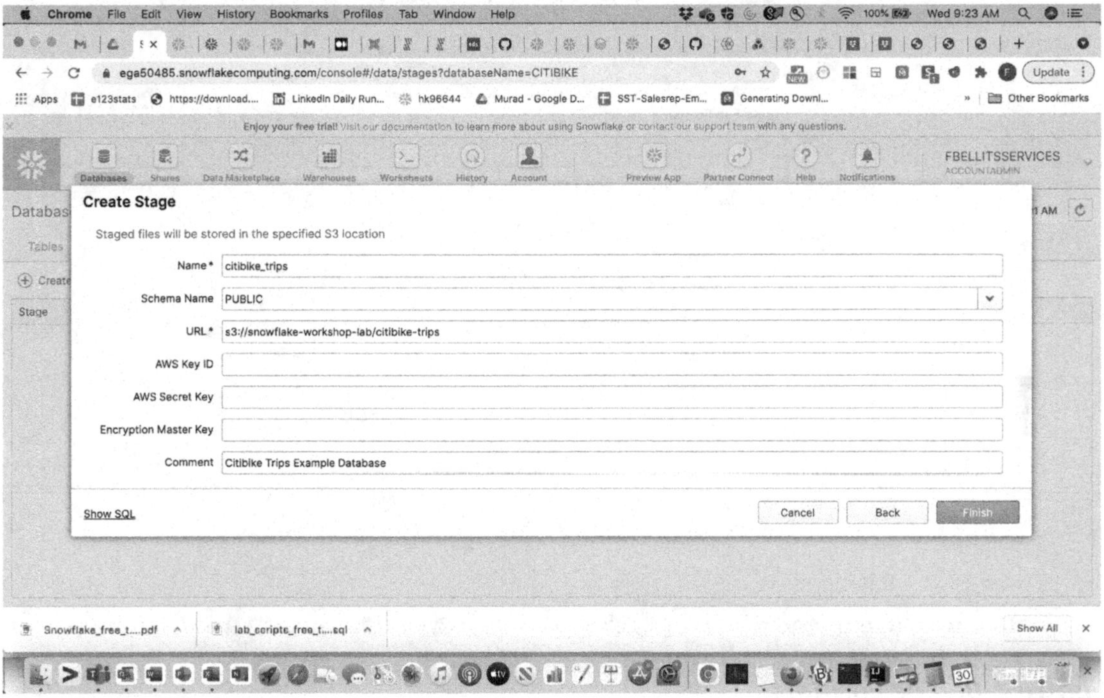

Figure 4-7. Creating AWS Stage

File Formats

File Formats similar to stages are specific to Snowflake. They are very similar though to relational database syntax that describes file type and format details so they can be loaded by bulk upload commands such as bcp (bulk copy program) for SQL Server, Oracle Loader for Oracle, nzload for Netezza, and FastLoad for Teradata.

File Formats simply describe the format of the file you are loading from a stage or directly from a COPY INTO statement. They can be dynamically defined within code as well. They provide the COPY INTO statement with the details of the file so it can be loaded correctly. You can Create, Clone, Edit, Drop, and Transfer Ownership of File Formats similar to other database objects.

File Formats have many different syntax options as shown in Figure 4-8. The figure is an example of the top of an empty form for creating a File Format that you will get when you hit Create File Format on the web interface.

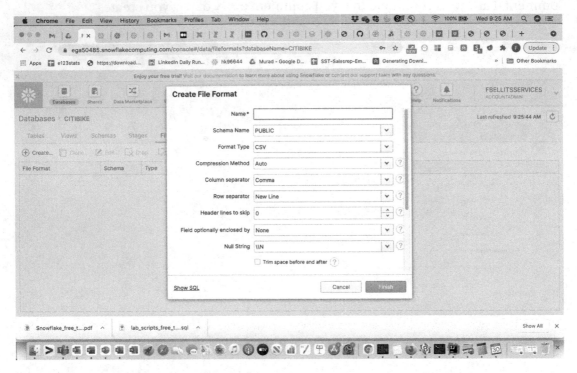

Figure 4-8. *File Formats*

The Create File Format form on the Snowflake interface provides several fields for you to fill out. The following are the field names and their descriptions:

- Name: Fill in the name of the File Format.

- Schema Name: Fill in the schema of the File Format that you are creating.

- Format Type: Available options are CSV (which DOES NOT specifically mean CSV, but really means Delimited File Type), JSON, XML, Avro, ORC, and Parquet.

- Compression Method: Available options are Auto, Gzip, Deflate, Raw Deflate, Bz2, Brotli, Zstd, and None.

- Column separator: Available options are Comma, Vertical Bar, Tab, None, and Other. You can add a custom column separator with the Other option, but it can ONLY be one character.

- Row separator: New Line, Carriage Return, None, or Other. You can add a custom row separator with the Other option.

- Header lines to skip: Enter the number of rows or lines (if any) that you want to skip.

- Field optionally enclosed by: None, Double Quote, or Single Quote. You use this to deal with the common delimited file issue of the delimiter being within the field or column. This encapsulates extra delimiters to enable the delimited data to load properly.

- Null String: Can be \\N, NULL, NUL, or Other. You can add a custom null string with the Other option.

- Trim space before and after checkbox: This enables the COPY INTO to trim white space before and after the field.

- Error on Column Count Mismatch: This is a key setting and typically for quality purposes is enabled. It identifies an error if the number of columns in the source does not match the number of columns in the destination.

- Escape Character: Backslash, None, or Other. You can add a custom Escape Character with the Other option. This is used to escape separators or special characters like single and double quotes.

- Escape Unenclosed Field: Backslash, None, or Other. You can add a custom Escape Unenclosed Field with the Other option.

- Date Format: Auto or Other. You can add a custom Date Format with the Other option.

- Timestamp Format: Auto or Other. You can add a custom Timestamp Format with the Other option.

- Comment: Enter a comment that describes the purpose or details of the File Format object.

Sequences

Sequences are another common database object in relational databases. Sequences allow users to increment and generate unique integer values for rows of data within tables. Figure 4-9 shows the view of the Create Sequence form.

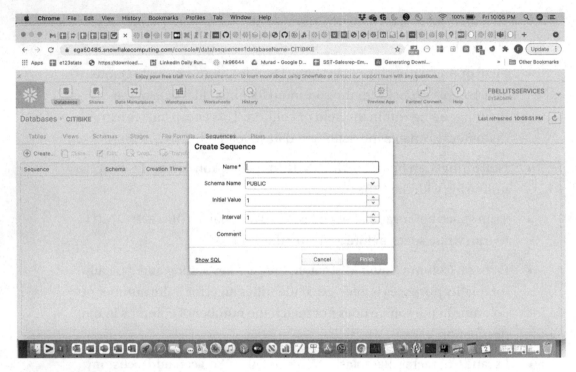

Figure 4-9. *Sequences*

You can Create, Clone, Edit, Drop, and Transfer Ownership of sequences within this web interface. You will notice once you select a sequence, similar to all the other database objects, you can grant permissions to the object as well on the right-hand side.

Pipes

Pipes or Snowpipes (the original name) within this interface are relatively new to the Classic Console. Pipes are one of the unique features of Snowflake that allow for the loading of continuous data pipelines as files are transferred into External Stages. Pipes can be set up to auto-ingest files into Snowflake based on the cloud provider event mechanisms. Figure 4-10 shows the first Snowflake web interface form for creating an initial pipe. Similar to all Snowflake objects, you can also create a pipe with data definition code.

Tip   Before you start creating your pipe, you will need a stage already created to use as the data source. Make sure to prepare the stage prior to creating the pipe. Also, if you plan to have a specific File Format you want to use in the pipe, then also create the File Format first before starting the pipe creation.

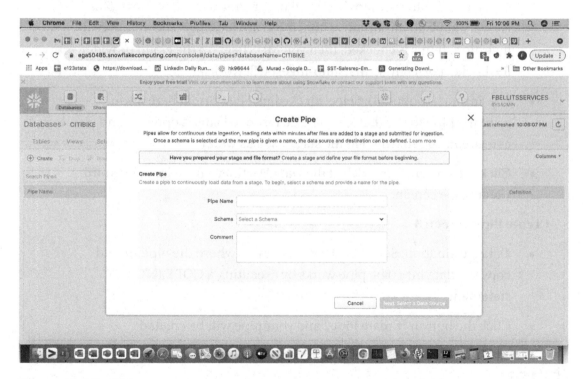

Figure 4-10. *Pipes*

Figure 4-10 shows the first of three dialogs that you fill out when creating a pipe. The three dialogs and the information you provide to each one are as follows:

Create Pipe: Screen 1

- Pipe Name: Enter the name you want to use for your pipe (Snowpipe).

- Schema: Enter the schema you want to create the pipe in.

- Comment: Enter a descriptive comment for what the pipe does.

- Click the button "Next: Select a Data Source" to go to the next Create Pipe screen.

Create Pipe: Screen 2 (Data Source for the Pipe)

- Stage: Select an existing stage name here from the dropdown. If you see "No options," you need to make sure you have access to a stage for the incoming data for the pipe.

- Enable Auto Ingest [this will ONLY be displayed if the stage has auto-ingest capabilities; otherwise, it will never show this checkbox]: This is a very key and important setting if you are looking to do automated ingestion. If this is checked, then you MUST set up the cloud provider file bucket correctly so that auto-ingest will work.

- File Formats (optional): Select an existing File Format here if you want the pipe to use that specific File Format. [This is optional and can already be specified in the stage itself as well.]

- Click the button "Next: Select the Data Destination" to go to the third Create Pipe screen.

Create Pipe Screen 3

- Data Destination: Select a schema and table where the pipe should copy the data into. The pipe works by executing a COPY INTO statement.

- Click the button "Create Pipe," and your pipe will be created.

Shares

Snowflake data shares are another unique and exciting feature that is part of the Snowflake Data Cloud. We will cover data shares, Data Exchanges, and Data Marketplaces in depth in Chapter 14. If you click the Shares icon and get the following message at the bottom of the screen as shown in Figure 4-11, then this means the current role you are using DOES NOT have access to create or view data shares:

Snowflake Data Sharing

Secure Shares enable you to consume data being shared with your organization and also provide data to others. Contact your account administrator to get access.

You need to switch to the ACCOUNTADMIN role, or a similar role provided by your organization, so you can view the screen in Figure 4-12, which shows a listing of data shares inbound. By default if you are using the ACCOUNTADMIN role, you will see two inbound data shares named ACCOUNT_USAGE and SAMPLE_DATA.

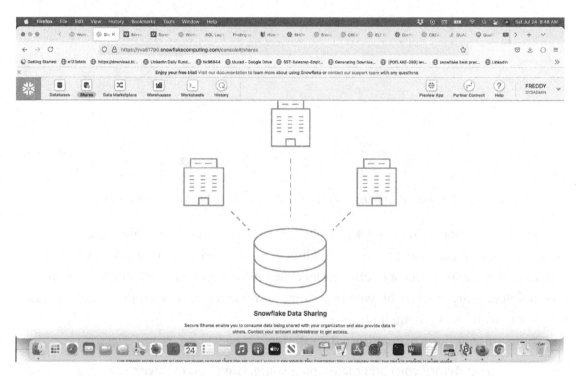

Figure 4-11. *Default View for Most Roles Besides ACCOUNTADMIN*

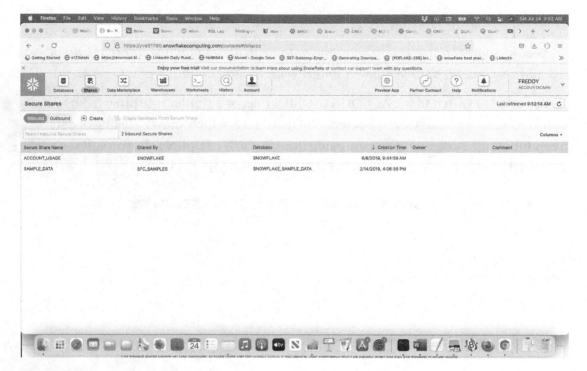

Figure 4-12. *Data Share Listings View for Roles with Data Share Access*

You will also notice in Figure 4-12 right next to the Inbound light-blue link is an Outbound link. If you click the Outbound link, you will see shares that you have created for external usage by other accounts. So let's dig into how you create an outbound or external data share to share between your own company, external suppliers, or any other external constituents you might have.

There are four parts to creating a data share:

Part 1: Before you can create a data share, you need to have the data properly prepared for sharing.

Part 2: Create the share itself.

Part 3: Review the secure share, preview tables, and validate secure views.

Part 4: Add consumers for the data share. This can be done by adding another account WITHIN that same region and giving it access to the share. If you try to add an account name not within the same exact region, you will get an error.

Figure 4-13 shows the first screen of creating a secure data share within the Snowflake Data Cloud.

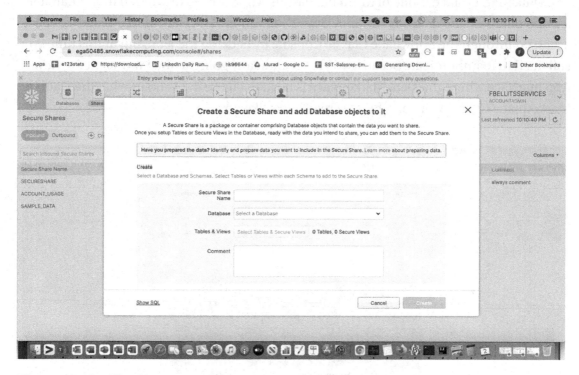

Figure 4-13. *Shares*

The form entries for a Snowflake data share are as follows:

- Secure Share Name: The name for the data share.

- Database: You select the database you plan to allow to be shared.

- Tables & Views: You select one or more tables and secure views (remember, regular views will not show up here).

You can add consumer accounts afterward.

Providing access to a share can be done either by adding existing accounts within the same exact region or by creating Reader Accounts. We will go into adding accounts in more depth in Chapter 14.

Data Marketplace

The Data Marketplace is one of the main reasons why Snowflake is much more than just a cloud database and is really the Snowflake Data Cloud. The Data Marketplace can be accessed through the Snowflake Classic Console by clicking the icon, but at this time you still need to reauthenticate because you are technically going to the Snowsight Preview App interface where the Data Marketplace is hosted. Figure 4-14 shows what the Data Marketplace Classic Console splash screen looks like now.

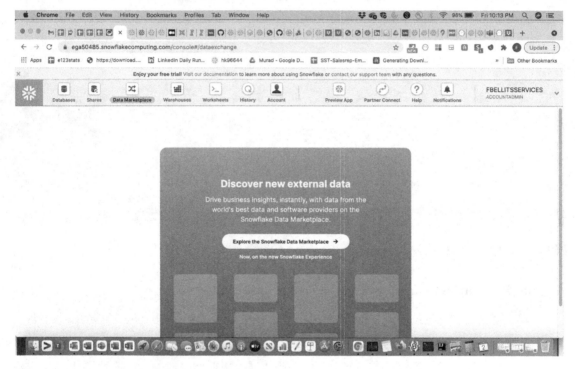

Figure 4-14. *Data Marketplace Splash Screen*

Once you click the "Explore the Snowflake Data Marketplace" button in Figure 4-14 and authenticate to the Snowflake Preview App, then you will be presented with the initial Data Marketplace Dashboard, which has listings of hundreds of Data Marketplace shares from hundreds of data providers. Figure 4-15 shows an example of the Data Marketplace initial interface. Also, remember that the Data Marketplace can be different from region to region right now. We also discuss the Snowflake Data Marketplace in depth in Chapter 14.

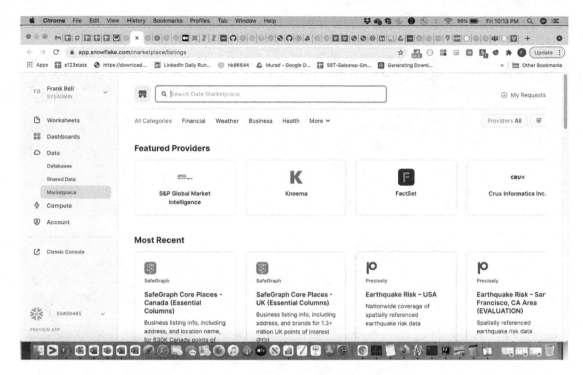

Figure 4-15. Data Marketplace Listings Initial Screen

In Figure 4-15 you can search for Data Marketplace shares based on provider name or share name or choose one of the Data Marketplace categories.

Warehouses (Named "Compute Warehouses" on the Preview App)

Warehouses are a new concept within the Snowflake Data Cloud, and they are technically virtual warehouses or really just pointers to compute resources within Snowflake and the cloud provider on top of which you are running Snowflake. You cannot execute SELECT, DELETE, UPDATE, INSERT, and MERGE statements without a warehouse assigned. If you are coming from a standard relational data warehousing background, then do NOT get confused by the naming convention Snowflake used here with "warehouses." Snowflake warehouses are completely virtual pointers and have nothing whatsoever to do with the storage of the data. The Snowflake Data Cloud has a clear separation of storage from compute, and warehouses are the compute part of the Snowflake Data Cloud.

Figure 4-16 shows the Create Warehouse form in the Classic Console on Snowflake.

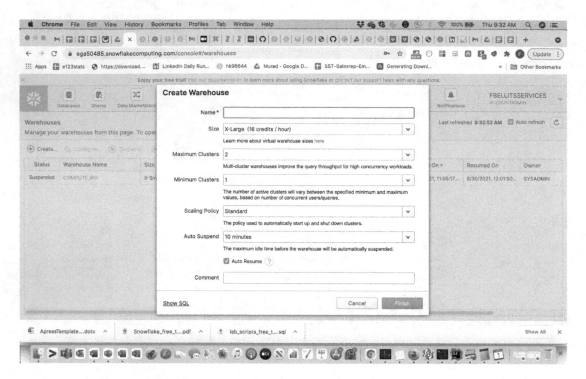

Figure 4-16. *Warehouses*

The form entries required for the Snowflake warehouse are

> Name: The name of the warehouse.

> Size: This is one of the most important selections and has
> MASSIVE impacts on the cost of your Snowflake Data Cloud. The
> cost difference between the smallest warehouse of XS (extrasmall)
> and the largest of 6XL is significant.

> Maximum Clusters: (You will not see this line if you have the
> Snowflake Standard version). This value can range from 1 to 10
> currently.

> Minimum Clusters: (You will not see this line if you have the
> Snowflake Standard version). This can range from 1 to 10
> currently.

> Scaling Policy: Economy or Standard. This impacts how fast an
> additional cluster turns on.

Auto Suspend: This is another incredibly important setting related to both cost and performance.

Comment: It is important to add your comment on what this warehouse is created for.

Worksheets

If you use the web interface exclusively, then this will be your main working area. The most important part of the worksheets in both the Classic Console and the Preview App is the context selection. This greatly impacts what your worksheet syntax must be. If you are running a command that is not fully qualified with the DatabaseName. SchemaName.FinalObjectName, then Snowflake will use what you have defined within the context for your database and schema. Also, unless you change the warehouse setting in the context, then Snowflake will also use that warehouse to run the worksheet workload details and charge credits based on the warehouse size and other settings. The warehouse will resume and suspend based on the warehouse Auto Suspend and Auto Resume settings as well, which have significant impacts on usage costs. Figure 4-17 shows a blank worksheet tab with the context dropdown settings shown.

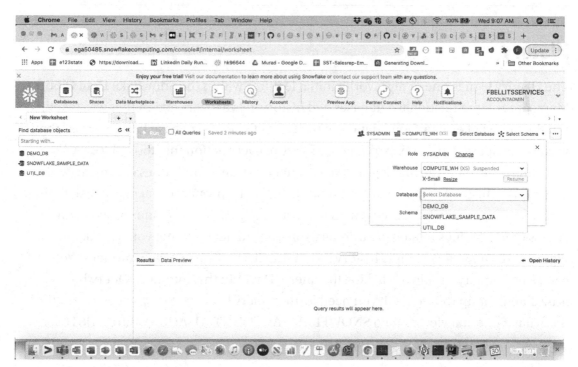

Figure 4-17. *Worksheets*

You will also notice on the Classic Console worksheet that the left navigation tree has both object exploration and search functionality. You can view the hierarchy of databases, schemas, and objects you have access to within the navigation tree. Remember that the navigation tree is dependent upon the current worksheet context that you are using. The worksheet has four main areas including the navigation tree, the worksheet input itself, the worksheet context (where you select Role, Warehouse, Database, and Schema), and the results pane. The results pane also has many options to view the query history or to copy or export data from the results. You can also access query profiles from here as well.

Tip   The role you are using when you create an object in the worksheet has a large impact on what roles and users can view or use the object. It is incredibly easy to mistakenly create an object (table, view, schema, and even database) in the worksheet while being in the role of ACCOUNTADMIN and forget to grant ownership or visibility to another role that needs access. When someone states they cannot find some object that you know has been created, the first thing to troubleshoot is what roles have access to the object.

History

The History icon is one of my favorite initial features within the Snowflake Data Cloud. Many preexisting relational databases had history logs of every action that was taken, but the Snowflake Data Cloud database really was the first database I came across that was so transparent about providing full history of every action that took place easily (assuming you have the privileges to see the actions). Snowflake does control access to the query history so that only the users who have the proper role access can actually view the queries they have access to in the history log. This is also one way you can access query profiles related to any query run on your account that you have access to. Figure 4-18 shows an example history log on the Classic Console web interface. You can select a query profile by clicking the query ID within this interface. Query history is available for up to 14 days. If you need further query history, then you can query the Snowflake internal view named SNOWFLAKE.ACCOUNT_USAGE.QUERY_HISTORY.

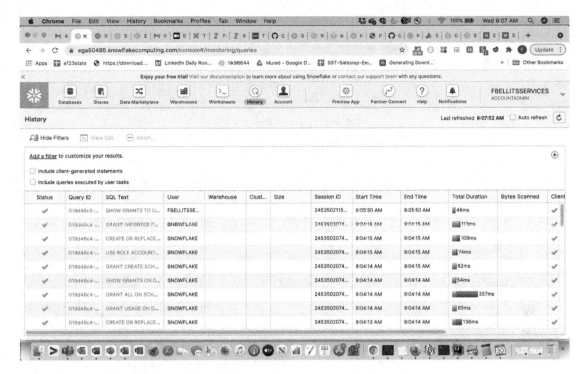

Figure 4-18. *History (Very Cool)*

Partner Connect

Partner Connect is another unique concept Snowflake came up with to showcase approved partner functionality and easily trial partner solutions. Figure 4-19 shows the initial top of the screen that you will see on the Partner Connect web interface. You can scroll down to find additional partners as well.

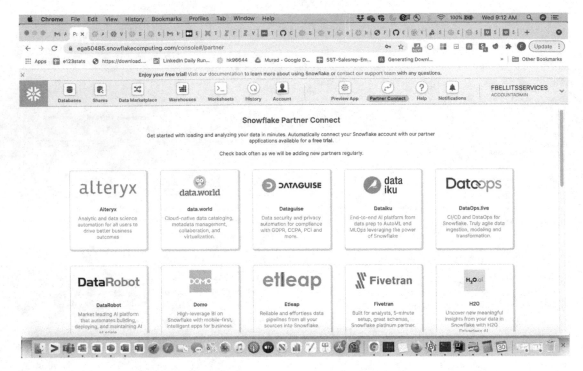

Figure 4-19. *Partner Connect*

Help

Help is a key component of any technical system. Snowflake has a simple but useful selection of help links to four main areas including the Snowflake documentation, Snowflake Community, Downloads, and the Help Panel. Most of these are straightforward and make it easy to find help details on Snowflake. The Downloads section provides links and installation details on how to set up Snowflake connectors and tools including the CLI client, JDBC driver, ODBC driver, Python components, Node.js driver, Spark connector, Go Snowflake Driver, and SnowCD (the Snowflake Connectivity Diagnostic Tool). Figure 4-20 shows the top-level options available when you click the Snowflake Classic Console Help icon.

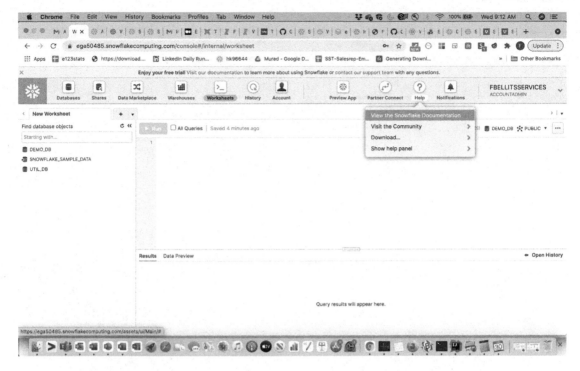

Figure 4-20. *Help Dropdown Screen*

Notifications

Notifications ONLY provide notifications to ACCOUNTADMINS currently. These notifications are important for monitoring spend on virtual warehouses and are related to resource monitors. They should be one of the first things that are set up at the beginning of gaining access to your Snowflake Data Cloud system. I recommend setting up BOTH web and email notifications at first and immediately configuring resource monitors, which we will discuss later. Figure 4-21 shows the view when you hover on the Notifications icon. You must click the settings link in Figure 4-21 to get to Figure 4-22 where you can set the notification options.

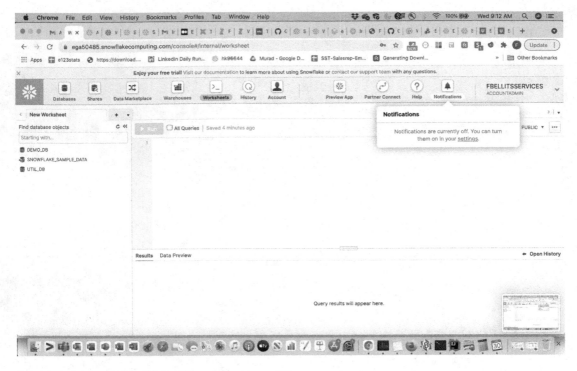

Figure 4-21. *Initial Notifications Icon View*

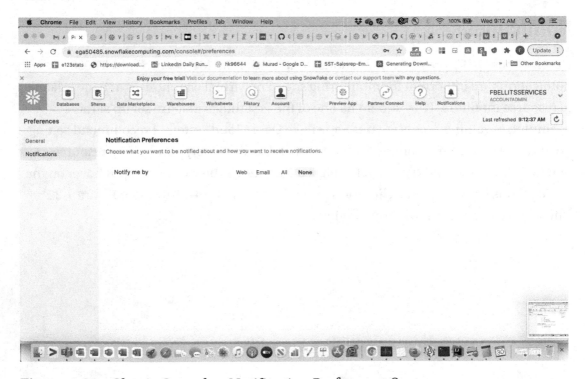

Figure 4-22. *Classic Console – Notification Preferences Screen*

Account

When you click the Account icon, you are taken to a web page with eight standard options including Usage, Billing, Users, Roles, Policies, Sessions, Resource Monitors, and Reader Accounts.

Tip   If you do not see the Account icon to the right of the History icon on the Classic Console, then you do not have rights to access the Account details with the current role you are in. Try changing the role to ACCOUNTADMIN if you have that access.

Usage

The Usage screen allows users to see their current daily compute spend and the spend breakdown by warehouses if the first rectangle with Warehouses and Credits Used is selected. Figure 4-23 shows what this warehouse spend view looks like on a new Snowflake Data Cloud account.

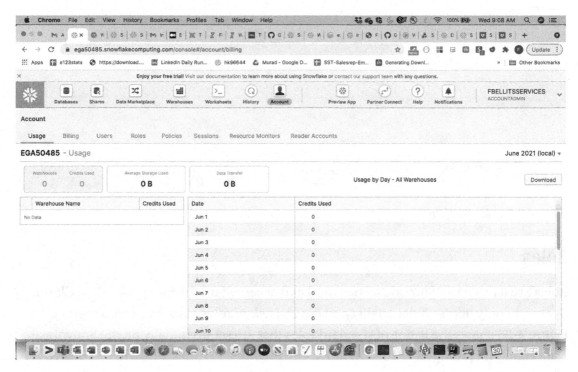

Figure 4-23. Usage

If you want to view storage usage, then you would click the rectangle just to the right of the "Warehouses and Credits Used" one, which is labeled "Average Storage Used." This view breaks down storage by database, stage, and Fail Safe. Finally, you can view any data transfer costs by clicking the rectangle to the right of this named "Data Transfer."

Billing

The Billing screen is where you can view and manage billing information for your account and add your credit card. This is typically used by on-demand accounts. Figure 4-24 displays an example of the Billing screen.

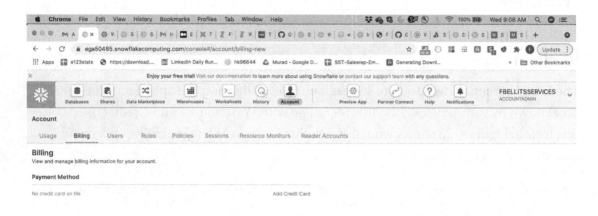

Figure 4-24. *Billing*

Roles

Roles are a key part of how data is secured and accessed on the Snowflake Data Cloud. Users can only access functionality and data based on roles associated with them. Figure 4-25 shows the listing of standard roles that are set up initially on the Snowflake Data Cloud including ACCOUNTADMIN, ORGADMIN, PUBLIC, SECURITYADMIN, SYSADMIN, and USERADMIN.

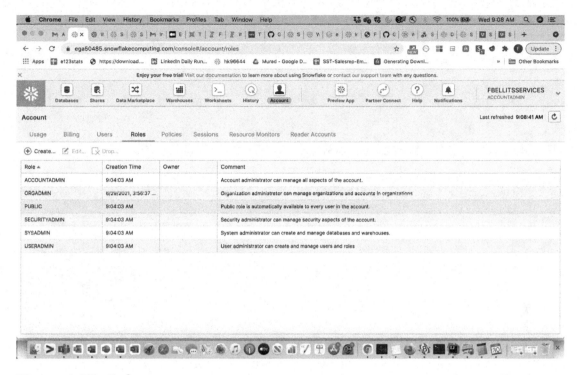

Figure 4-25. *Roles*

Policies

Policies in Snowflake are used for network security. Figure 4-26 shows an example of the Create Network Policy form. You can choose what IP addresses you want to allow and those you want to block. You can have multiple policies as well.

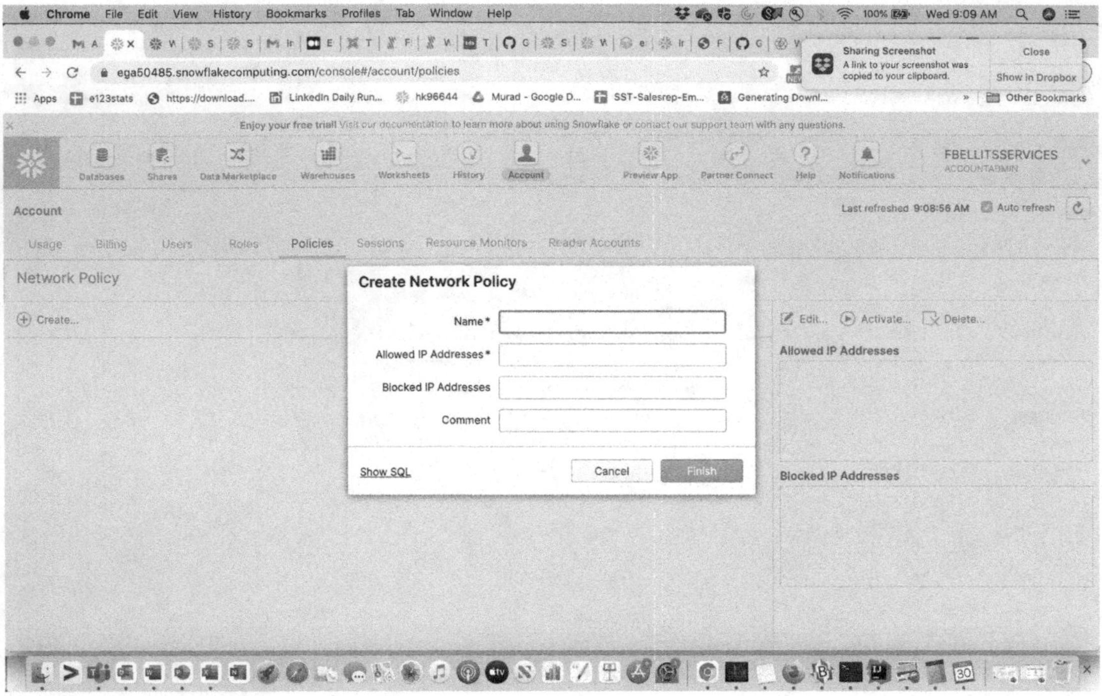

Figure 4-26. Policies

Sessions

The Sessions link under the Account area allows you to see currently active sessions in your Snowflake Data Cloud account. The session listings include details on User Name, Session ID, Open, Start Time, End Time, Duration, Expiration Time, Client Driver, Client Net Address, and Authentication Method. Figure 4-27 shows the session display with two active sessions.

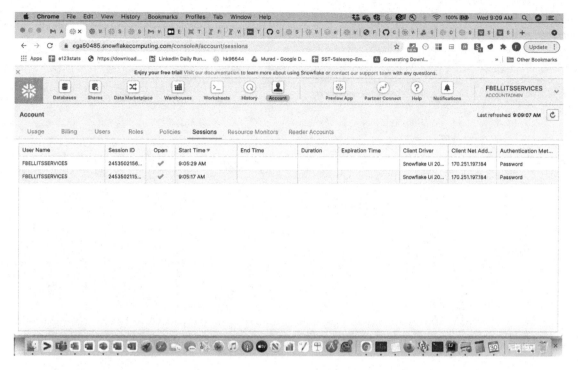

Figure 4-27. *Sessions*

Resource Monitors

Resource monitors are one of the most important and often neglected features of the Snowflake Data Cloud. When on-prem users transition to the Snowflake Data Cloud, they often love the flexibility, ease of use, and unlimited scalability. Database practitioners who used on-prem databases must make a shift to using tools to monitor cloud consumption, storage, and other costs. Most on-prem databases just had fixed prices for their systems, but cloud databases typically have consumption pricing. The ONLY tool currently available by default on the Snowflake Data Cloud that monitors consumption and shuts off warehouses are resource monitors.

Resource monitors can be used in two main ways to either notify Account Admins of hitting usage thresholds or to shut down the warehouse. One or multiple notifications can be set up. When a warehouse hits resource monitor thresholds, then it can either be suspended immediately or after the current query that went over the threshold finishes.

Figure 4-28 is an example of the Create Resource Monitor form. You will notice you can set up resource monitors for one warehouse, multiple warehouses, and overall accounts. Also, warehouses currently can be set for monitoring intervals of Daily, Weekly, or Monthly.

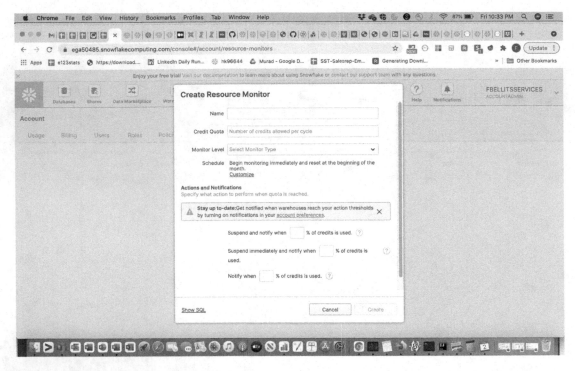

Figure 4-28. *Resource Monitors*

Reader Accounts

Reader Accounts are a type of account Snowflake created to allow data share providers a way to share their data with data consumers who do not have an existing Snowflake account. The main difference between a Reader Account and a normal Snowflake account is that the Reader Account consumption is paid for by the data provider. Also, the data provider is the one who creates the Reader Account. Figure 4-29 shows the standard form where you can create a Reader Account assuming you are currently using a role like ACCOUNTADMIN.

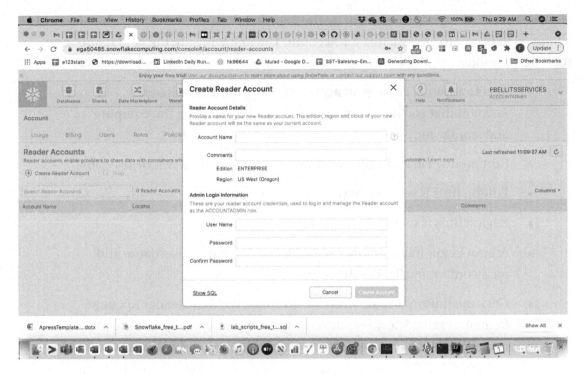

Figure 4-29. *Create Reader Accounts*

Reader Accounts make it incredibly easy for a data provider to share data to their customers who do not have an existing Snowflake account. You can easily set them up and share data within minutes. While they provide great ease of use and power, you must remember these are full Snowflake accounts that the primary account is paying the Snowflake costs for. You should never provide the administrative username and password to the reader party unless you have some agreement in place for the costs and administrative responsibilities. Otherwise, the best practice is to go in as the administrator username you just created and create roles with typically limited access specially to creating and using warehouses. We also recommend setting up resource monitors immediately so you can track all the usage effectively on the Reader Account.

The Create Reader Account form on the Snowflake Classic Console shows the fields you need to fill out to create a Reader Account. The following are the field names and their descriptions:

- Account Name: Provide an account name for the new Reader Account that you [the primary account] will be paying the compute and storage for.

- Comments: Fill in details of what the Reader Account will be used for.

- Edition [this is set to what your current account is]

- Region [this is set to what your current account is]

- Admin Login Information: Create an administrative username and password for the Reader Account.

- Click the button "Create Account" to create the new Reader Account. Remember that Snowflake accounts now typically take a minute to get fully created. If you go to the account URL too early, you may see a 400 error.

Once you are finished creating the Reader Account, make sure to set up the security as mentioned previously before giving access to the reader data consumers. Remember that all the usage performed on the Reader Account will be paid for by the primary account, which created the Reader Account. In Figure 4-30 you can see what the initial screen will look like for the new Reader Account.

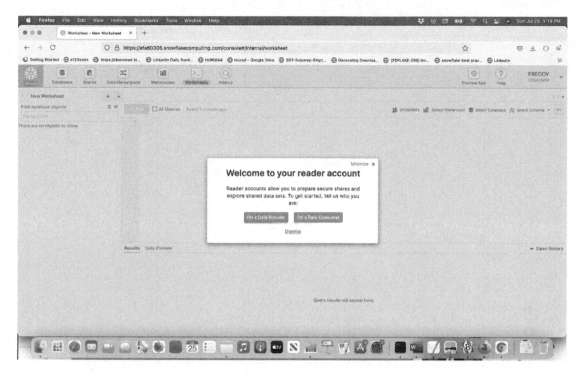

Figure 4-30. *New Reader Account*

Summary

The Snowflake Classic Console has been the standard web interface for using the Snowflake database since it launched. It still provides an excellent initial interface for executing queries on Snowflake and also managing the Snowflake Data Cloud. The new Snowsight interface, which we cover in Chapter 5, does provide some amazing new features related to autosuggestion, autocompletion, and sharing of worksheets and dashboards. Currently you also must access Snowsight (Preview App) through the Snowflake Classic Console. There is no current date scheduled for removing the Classic Console, so all of us veterans used to using it can continue to do so for a while.

Snowflake Web Interface: Preview App (Snowsight)

In this chapter, we will cover Snowflake's Preview App (aka Snowsight) web interface and all the functionality within it. Snowsight is Snowflake's new web interface that allows you to interact with the Snowflake Data Cloud and perform data operations and Snowflake administration tasks from a visual web interface. As I write this, Snowflake has re-released their Snowsight interface preview and added a lot more functionality to it. We will cover all the details in this chapter. As of this writing, you will need to access the Preview App (Snowsight) from the Classic Console. Also, someone with the ACCOUNTADMIN role must first log in to the Preview App and enable it for other users of the account as we show later in Figure 5-3.

Web Interface: Preview App (Snowsight) Main Overview

The Preview App interface also named Snowsight is Snowflake's planned future full web interface for the Snowflake Data Cloud. Currently it is still in preview and contains some useful new features. As of this writing though, Snowsight is still missing some key features to make it fully usable for all functionality the Classic Console performed. Snowsight was created by Snowflake to improve overall data collaboration and ease of use of the Snowflake Data Cloud. Snowsight also provides a much more transparent interface into what functionality is available in Snowflake than the initial Snowflake Classic Console. In the Classic Console, you had no idea that all the features in the Account area even existed unless you had the ACCOUNTADMIN role or equivalent. While you may not be able to create or view all Account objects, you can at least see what is possible within the Snowsight interface.

© Frank Bell, Raj Chirumamilla, Bhaskar B. Joshi, Bjorn Lindstrom, Ruchi Soni, Sameer Videkar 2022
F. Bell et al., *Snowflake Essentials*, https://doi.org/10.1007/978-1-4842-7316-6_5

Key differentiated features that Snowsight has that are not on the Classic Console are the following:

>*The ability to search across worksheets.

>*Visualization functionality including charts, tiles, and dashboards.

>*The ability to share dashboards and worksheets to other users in your Snowflake account.

>*The removal of the Classic Console bug where if you opened the same worksheet on two different browsers, the Classic Console could not keep any changes in sync, and you had to overwrite the worksheet or create a new one.

>*Autocomplete for objects in the worksheets.

>*The ability to more easily navigate to other accounts in your organization.

>*Visibility of user-created objects such as procedures, user-defined functions, and tasks. (Previously, on the Classic Console, you could never view them in the web interface.)

>*Full access to the Data Marketplace (you can get to this from the Classic Console, but it is sending you to a part of the Preview App.)

>*Centralized object creation capability from the web interface. Previously you had to navigate to a specific object listing to create it.

The six main left-hand navigation areas of Snowsight are

1. Worksheets

2. Dashboards

3. Data

4. Compute

5. Account

6. Organization [you will ONLY see this in the navigation if you have an account that is part of an overall organization]

Key Classic Console features that are not still available as of this book writing in Snowsight are

1. Create or view File Formats.

Navigation Tip   Whenever you are in Snowsight and do not have the left-hand navigation present, you can always click Home in the upper left of your screen to get your left-hand navigation back.

Initial Preview App (Snowsight) Login

The first time you log in to the "Preview App" from the Classic Console, it can be slightly confusing because even though you were authenticated on the Classic Console, it will still make you reauthenticate via the following two screens in Figures 5-1 and 5-2. These are necessary steps currently for you to initially log in to the "Preview App" (Snowsight). You will need to click the button in Figure 5-1 to get to the sign-in screen for Snowsight in Figure 5-2.

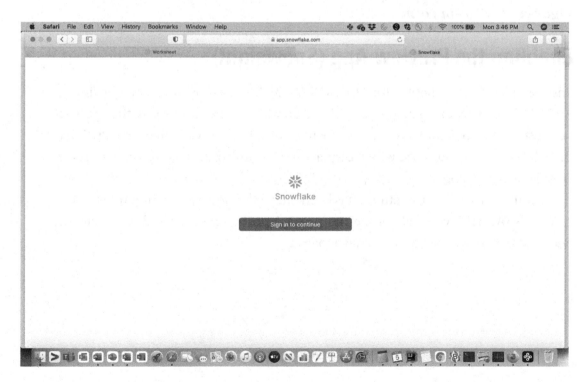

Figure 5-1. *Sign-In Button*

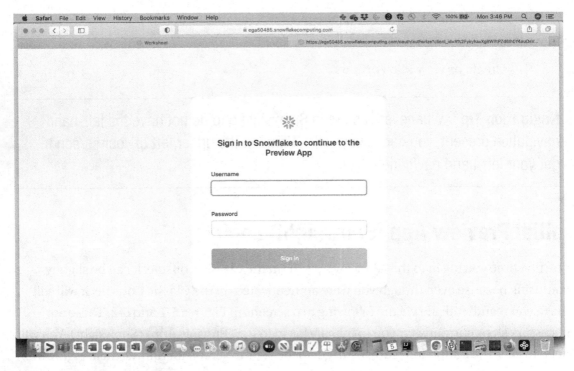

Figure 5-2. Sign-In Form

Enabling the Preview App (Snowsight)

Snowsight is initially enabled for ACCOUNTADMIN roles only. If you are the first
ACCOUNTADMIN to log in to the Preview App with the previous steps, then you will
be presented with Figure 5-3 initially or when you click the Worksheets navigation in
the left-hand navigation bar, which displays the "Ready to start using Worksheets and
Dashboards?" dialog.

   If you are accessing this after it has been enabled or from a different role than
ACCOUNTADMIN, you will not see Figure 5-3 since the account would have already
been enabled for worksheets and dashboards.

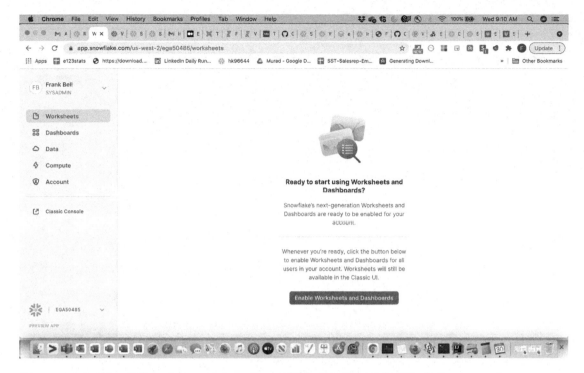

Figure 5-3. *Enabling Snowsight – Just Click the "Enable Worksheets and Dashboards" Button*

Worksheets

The Worksheets area in Figure 5-4 is the main querying and coding section of the Preview App (Snowsight) interface similar to the Classic Console Worksheets. The Preview App (Snowsight) Worksheets though have a much different display for multiple worksheets where worksheets are listed within the main panel of the application vs. worksheet tabs toward the top of the interface. The Preview App also has more advanced functionality with autocomplete and worksheet sharing available. In Figure 5-4 you can see the initial Worksheets interface after it is enabled. At this point you can either create a new worksheet to work from by clicking the upper right "+ Worksheet" button, or you have the option to import worksheets with the "Import Worksheets" button, which will import worksheets from the Classic Console that you had created previously.

Figure 5-4. *Snowsight Worksheets Initial View*

Let's cover both worksheet interface options of importing worksheets and creating a
new worksheet.

Figure 5-5 shows the dialog that comes up if you decide to import some of your
previous work from the Classic Console worksheets. It will list the number of worksheets
it can migrate over. If you are not ready to use the Preview App extensively, I recommend
cancelling and waiting until later. The issue is if you are still using the Classic Console
mostly, then it can get confusing if you import many worksheets that you have created
previously and do not switch over to using the Snowsight interface most of the time. The
Classic Console and the Preview App (Snowsight) at this time are completely different
environments and do not share worksheets OR worksheet changes between them. It can
become very easy to change worksheets either in the Classic Console or Preview App
(Snowsight) and not have them changed in the other interface creating discrepancies
and challenges for you. Our best practice approach is to use some type of source
code repository such as Git to share all your SQL code. Then you can navigate these
environments knowing you have one source of SQL query or object source code truth.

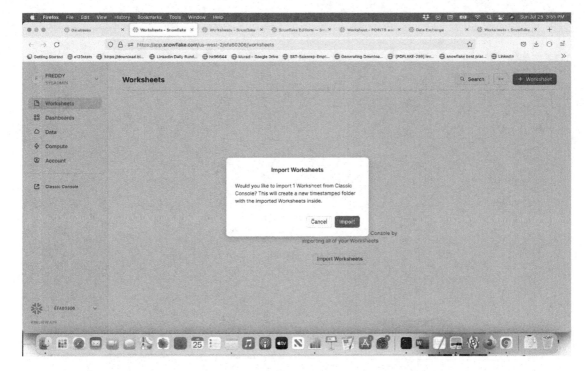

Figure 5-5. *Initial Import Worksheets Message*

If you just want to create a new worksheet to get started from, then just click the upper-right "+ Worksheet" button, and you will come to a screen that looks like Figure 5-6.

Figure 5-6. *Initial Worksheet Detailed Screen*

A new empty worksheet is packed with all sorts of navigation features, so let's step through them one by one since they are highly important to your productivity if you choose to write queries and code from the new Snowsight worksheet interface.

The worksheet detailed interface has the following navigation features going clockwise from the upper-left Home icon:

- Home Icon: At first it can be confusing to all who are familiar with left-hand pane navigation elements (Worksheets, Dashboards, Data, Compute, etc.), as they create a new worksheet. The Home icon is the way to get back to them. It immediately takes you to a list of worksheets and also returns that friendly left-hand pane of navigation.

- Worksheet Name: The date and time are initially filled in as the worksheet name in the upper left of the screen just to the right of the Home icon. We recommend naming your worksheet something specific immediately. You will also notice when you go to change the worksheet name, there is a lot of other functionality in that dropdown including the capability to "Move to" – move the worksheet to both folders and dashboards – or create a new folder or new dashboard to move it to. You can also "Duplicate" the worksheet, "Format the query," "Delete" the worksheet, or view shortcuts.

- +: This just creates another blank worksheet.

- Context Section for Roles and Warehouses: Notice the change from the Classic Console where this worksheet context had four items and included the database and schema as well. Now that is separated. This interface at first will most likely display your role of ACCOUNTADMIN, SYSADMIN, PUBLIC, etc. and your default warehouse. When you click it, you can select different roles and warehouses that are available to those roles.

- Share Button: This new exciting feature to share worksheets we will cover in the following.

- > (Execute): Executes the query in the worksheet.

- Worksheet Status and Versions. At first your worksheet will say Draft right below the execution arrow. Once you have worked with a worksheet for a while, a dropdown is displayed, and you can use it to select different versions of the worksheet (history of the worksheet changes).

- Main Worksheet – Database and Schema Context Selection.

- Main Worksheet: This is where you can create your queries, DML, procedures, functions, etc. Notice it has numbering to the left as well for line numbers. Autosuggest works within this pane.

- Main Worksheet Bottom Left – Objects: Clicking the "Objects" button hides the "Databases" and "Pinned" object sections on the left. Clicking it again makes them reappear.

- Main Worksheet Bottom Left – Query: Clicking the "Query" button makes the initial query hidden and then displays all the results if there are any.

- Main Worksheet Bottom or Left – Results: This is only displayed when there are results from your query. You can click this to go from a chart back to tabular results.

- Main Worksheet Bottom Left – Chart: This is only displayed when there are results from your query. You can click it to display a chart instead of tabular results. You will also notice that there are many selections for Chart, and we will cover those later, but at a high level right now you can choose a type of chart (Line, Bar, Scatter, Heatgrid, Scorecard), columns used in the chart, and appearance specifics.

- Main Worksheet Bottom Right – Search Icon: Allows you to search through the results.

- Main Worksheet Bottom Right – Download Icon: You can download the results to a file.

- Main Worksheet Bottom Right – Split Panel Icon: [need more details]

- Databases: If the role you have selected has access to databases, then they are listed here in a similar tree navigation pane. You can navigate to databases and schemas within as well as the objects of tables, views, stages, data pipelines, functions, and procedures within the schema. You can also click any of the objects and get different actions you can perform. For databases and schemas, you will see "Place Name in SQL" or "Set Worksheet Context." For tables and views, you will see "Show Details," "Place Name in SQL," and "Pin." For stages, data pipelines, functions, and procedures, you will see options for "Place Name in SQL" and "Pin."

- Pinned: You can pin objects to have them displayed in this panel.

- Filters: You can add filters, which will be displayed when you click the Filter icon.

You will also notice when you start working with the worksheets inside the Preview App, one of the best enhancements is autosuggestion and autocompletion of objects within the SQL queries. Figure 5-7 shows an example of how autocomplete works in Snowsight and shows how it not only autosuggests object names like table names but it also suggests SQL keywords. If you want any of the suggestions, just select them and bam, you saved a lot of keystrokes!

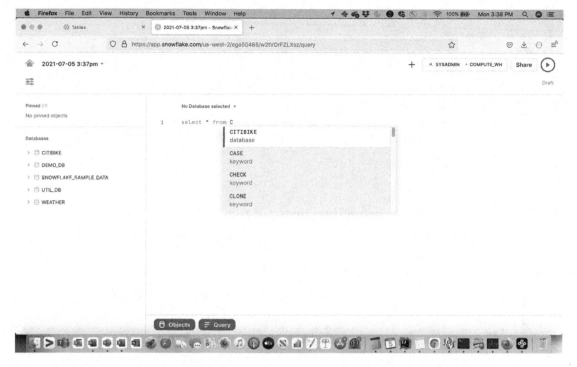

Figure 5-7. *Snowsight Worksheets Enhancement – Object Autocomplete*

I hope the preceding details gave you a very comprehensive overview of all the navigation capabilities when you are within a worksheet. Now let's click the Home icon in the upper left and go back to the main navigation pane with the worksheet listings. Figure 5-8 shows an example of what the screen looks like if you have folders. This specifically shows folders created when I imported Classic Console worksheets on multiple occasions.

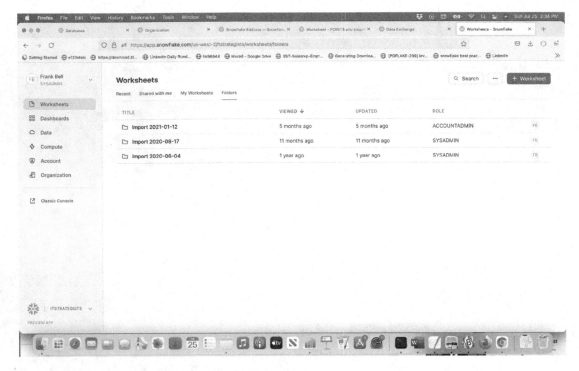

Figure 5-8. *Snowsight Worksheet Listings Screen*

You can view your worksheet listings by four different filters including Recent, Shared with me, My Worksheets, and Folders. We covered the navigation within the detailed worksheet screen, but you will notice in Figure 5-8 that as soon as you navigate to the worksheet listings where the left-hand navigation reappears, there are additional navigation elements for Worksheets specifically besides the filtered navigation:

- Search Button – Upper Right: You can finally search across all your worksheets very easily by entering your search details.

- … Button: If you click the … button, you will be provided with three dropdown options of creating a New Folder, being able to again Import Worksheets, and Manage Filters.

- + Worksheet: You can create a new worksheet here as we demonstrated previously.

- Filters: You can add filters, which will be displayed when you click the Filter icon.

- Main Worksheet Listings: You can click any of the worksheet listings and then go back to the detailed worksheet screen.

Initial Preview App (Snowsight) Role Selection

Like the Classic Console, your current role selection determines what you have visible to you on Snowsight as well as what is active on the various interfaces. Remember this as you navigate through Snowsight. Figure 5-9 shows how you can change the context of your Snowsight interface based on what role you have currently selected. This is the way it works ONLY when the left-hand navigation roles are assigned to the username you are logged in as.

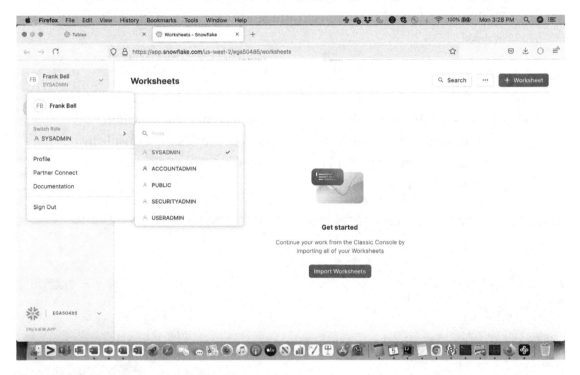

Figure 5-9. *Snowsight Role Selection*

Overall Navigation Pane

Before we jump into dashboards, let's cover all the generic navigation areas that are displayed when you have the left-hand navigation screen visible and see how they are different than when you are in the details of the worksheet and other detailed navigations. Both Figures 5-8 and 5-11, which show the worksheet listings and the dashboard listings, respectively, also give us a screen where we can explain all the generic navigation elements starting from your initials (or potentially your profile image) in the upper-left part of the screen:

- Picture or Initials and Profile Details Dropdown: If you click the upper left with your name and initials along with your currently selected role, then it provides a set of dropdown selections including your role selection we just covered previously and Profile, Partner Connect, Documentation, and Sign Out links. Figure 5-10 shows what a profile screen looks like.

- Six Main Areas of Left-Hand Navigation: This includes Worksheets, Dashboards, Data, Compute, Account, and Organization [if you have access].

- Classic Console Link: This allows you to go back to the Classic Console.

- Bottom-Left Account Selection: This section allows you to switch accounts much more easily than you could in the Classic Console by clicking the down arrow and just selecting what Snowflake account you want to work in.

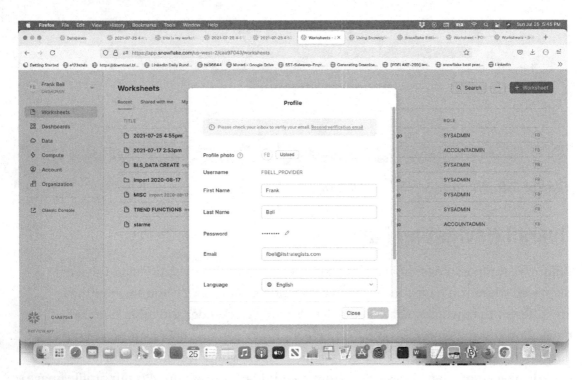

Figure 5-10. *Snowsight – Profile Screen Example*

You can do the following in the profile screen:

- Verify your email.

- Add a profile photo.

- View username.

- Change/enter first name.

- Change/enter last name.

- Change your password.

- Change/update your email.

- Change/update your language.

- Enroll in multi-factor authentication (MFA).

Dashboards

Dashboards are new visualization functionality within Snowsight. Remember that the dashboard functionality and navigation is only available within the Snowflake Preview App (Snowsight) interface. Snowflake added some neat visualization features such as dashboards, tiles, and visual chart functionality to allow you to visualize data within the Snowsight interface. Queries, worksheets, worksheet results, worksheet charts, and worksheet tiles are all really the same interrelated object. Worksheets are the working area of the query, and the worksheet result is the tabular form of the executed query result. The worksheet chart is the chart visualization of the query result. A tile is just a manifestation of those results that is placed into a panel within a dashboard.

Let's get going with dashboards. The first time you come to the dashboard listings, it will present you with Figure 5-11.

Figure 5-11. *Dashboard Initial Screen*

After you create your first dashboard, you will never see that screen again. You will instead see a variation of Figure 5-12 dashboard listings. Figure 5-12 shows an example dashboard listings page, which lists out two dashboards in the Recent view.

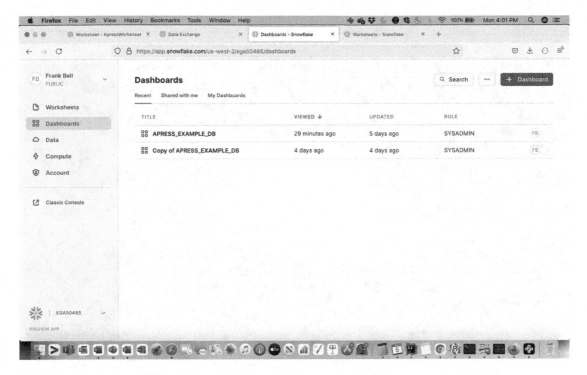

Figure 5-12. *Dashboard Listings*

You can view your dashboard listings by three different filters including Recent, Shared with me, and My Dashboards. Let's cover other Dashboard navigation functionality:

- Search Button [Dashboards] – Upper Right: You can search across all documents including dashboards and worksheets very easily by entering your search details.

- … Button [Dashboards]: If you click the … button, you will be provided with three dropdown options of Managed Filters.

- + Dashboard: You can create a new dashboard here.

For you to create a functional dashboard, you need at least one tile based on a query within a worksheet. Figure 5-13 displays the New Dashboard dialog that opens up after you click the + Dashboard button. We gave our dashboard a name of CITIBIKE. Now just click the button "Create Dashboard" to finish creating your dashboard shell so you can add a tile to it.

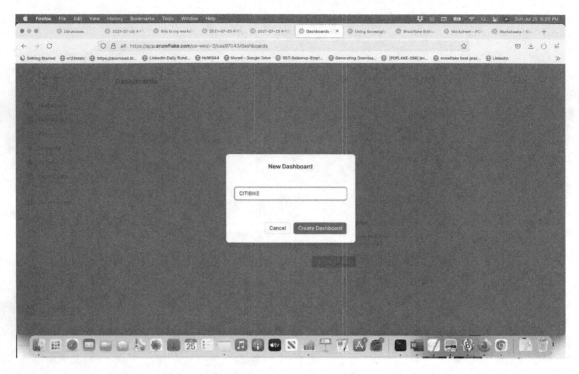

Figure 5-13. *New Dashboard Dialog*

Now that you have a dashboard shell, let's add a tile to it. Figure 5-14 displays the Add Tiles page you need to use to add a new tile. You can add it from the bottom center or the upper-left + sign.

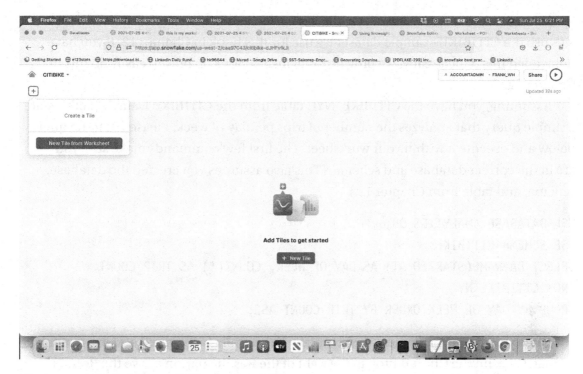

Figure 5-14. *Adding a Tile to a Dashboard*

Technically a tile really is the result of a worksheet. [I know, it is kind of confusing, but that's the way it works now.]

If you already have a query that you ran in a worksheet that produced either a tabular or chart result, you can reuse it for your tile here. Once you do that and come back to the dashboard, you have ... on the upper right of any tile within the dashboard detailed section. You can click that to get the following options:

- View table [which takes you back to the results pane].

- Edit query [which takes you back to the query itself, with the results pane below it].

- Duplicate tile [which will immediately duplicate the tile and split the screen by adding a panel; if you have one panel, it will become two panels].

- Unplace tile [this removes it from the current dashboard].

- Delete [this would delete the tile or worksheet permanently and cannot be undone].

If you did not have a worksheet with a result you can use in the dashboard then let's use the CITIBIKE example dataset we use for this book in Chapter 12. If you have not loaded this data already, we provide exact instructions on how to easily do this in Chapter 12.

Assuming, you have the CITIBIKE_NYC table from the CITIBIKE Data Set, let's create a simple query that analyzes the number of trips per day of week. Please enter the query below and execute it within your worksheet. The first few commands make sure you are in the correct database and schema. (This also assumes you created the database, schema, and table from Chapter 12.)

```
USE DATABASE ANALYTICS_DB;
USE SCHEMA CITIBIKE;
SELECT DAYNAME(STARTED_AT) AS DAY_OF_WEEK, COUNT(*) AS TRIP_COUNT
FROM CITIBIKE_NYC
GROUP BY DAY_OF_WEEK ORDER BY TRIP_COUNT ASC;
```

Once you execute this query it will initially has Results highlighted and it's in tabular format. Click on the Chart Button to the right of the Results button above the Results screen area. It will turn the tabular results to a chart. First, change the chart type over to the right to the Bar type. This should change the first column name in light green to TRIP_COUNT and SUM. If not than change the first column setting to TRIP_COUNT and SUM if you need to under the Data area on the right. Then change the next column from TRIP_COUNT to DAY_OF_WEEK. Finally, look for the Appearance section below this (Sometimes its hidden and you need to scroll) and change Orientation and this will change it from X-Axis to Y-Axis and then your screen should look similar to Figure 5-15 below which shows a chart in a Worksheet.

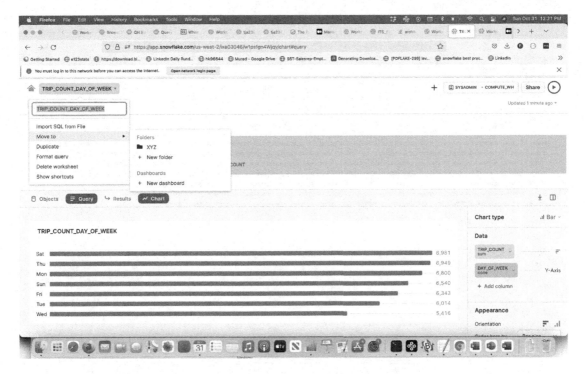

Figure 5-15. *Worksheet with Chart from Citi Bike Data Set*

You will notice in Figure 5-15, we also named the Worksheet in the upper left corner with the name TRIP_COUNT_DAY_OF_WEEK. Also, in the upper left to the right of that new name of the Worksheet and the Tile [They are technically the same thing] you will see a "Move To" selection where you can move this worksheet to a New Dashboard where the +Dashboard is. Enter a name of "CITIBIKE DASHBOARD" for your new Snowflake Dashboard. Also, if you need to make sure to navigate back to this exact CITIBIKE DASHBOARD. If you repeat these actions above then you should have a Dashboard visualization similar to Figure 5-16 below.

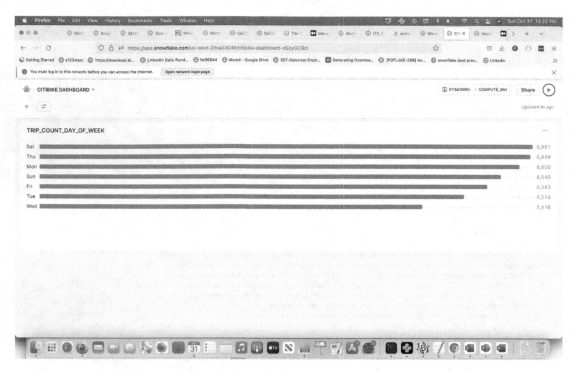

Figure 5-16. *Dashboard with Worksheet Chart from CITIBIKE Data Set*

Congratulations! You have created your first Dashboard with one tile in it. Notice as well that the Dashboard Detailed screen is very similar to the Worksheet screen except it has some additional specific Navigation elements as follows:

- Home icon. At first it can be confusing to those used to the left-hand pane navigation elements (Worksheets, Dashboards, Data, Compute, etc.) in the Classic Console, because currently these left side navigations disappear when you create a new worksheet or when you create a new dashboard. Similarly, whenever you currently go into a worksheet or dashboard the left side navigation disappears. The Home icon is the way to get back to them. It immediately takes you to a list of Worksheets and also returns that friendly left-hand pane of navigation.

- Dashboard Name Dropdown: This dropdown can be used to either duplicate or delete a dashboard.

- Context Section for Roles and Warehouses: Notice the change from the Classic Console where this worksheet context had four items and included the database and schema as well. Now that is

separated. This interface at first will most likely display your role
of ACCOUNTADMIN, SYSADMIN, PUBLIC, etc. and your default
warehouse. When you click it, you can select different roles and
warehouses that are available to those roles.

- Share Button: This new exciting feature to share dashboards we will
  cover in the following.

- Dashboard Tiles Displayed over Most of the Interface.

- +: This allows you to actually move existing tiles into the dashboard
  or create a "New Tile from Worksheet."

Now you have created your example dashboard with CITIBIKE example data.
Exciting! Before we go to the next major navigation area, let's show how easily you
can share dashboards and worksheets. Click the Share button just to the left of the
execution arrow. Figure 5-17 shows the sharing message for dashboards. You can easily
select someone from your team on the same account to share the dashboard with. The
worksheet sharing is the same.

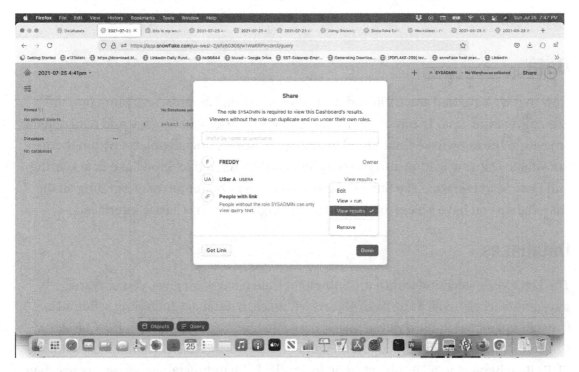

Figure 5-17. *Dashboard Share*

Notice the details around how you can share the dashboard as well. You have a lot of sharing capabilities here and can share the dashboard differently depending on each person, or you can also set privileges just on the dashboard link itself. When you share the results with a named user with a role to access this, then you can select the following options:

- Edit

- View + run

- View results

- Remove

When you share the results just with a link to them, then you can select the following options:

- View + run

- View results

- Cannot view

Be careful here because by default Snowflake has made people with link be able to view results.

Data

Now that we have had some fun with worksheets and dashboards, let's jump into the Data navigation area. The Snowsight Data area is where all data storage and data objects are within the Snowflake Data Cloud including databases and all objects within them. Snowflake Data navigation also includes Shared Data related to Snowflake data sharing and Data Marketplaces. The Snowsight interface does a better job of integrating all the shared data you have access to incoming and outgoing vs. the Classic interface.

Databases

The Databases selection within the Snowsight Data navigation gives you easy tree navigation access to all Snowflake objects within each database including schemas (by schema name), tables, views, stages, data pipelines (named pipes in the Classic Console), functions, and procedures. Figure 5-18 shows the CITIBIKE database with the PUBLIC schema and all its objects that are available including tables, views, stages, data pipelines, functions, and procedures.

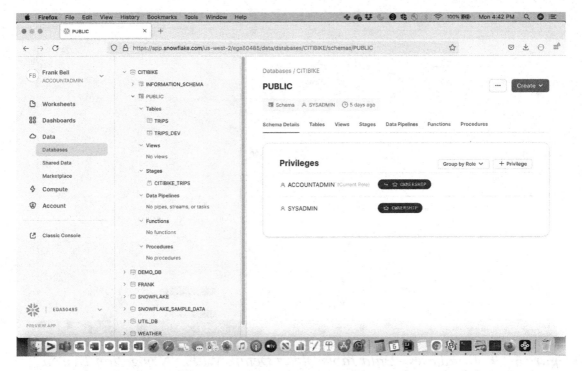

Figure 5-18. *Databases Tree Navigation on Preview App (Snowsight)*

The new interface has better visualization around not just the tree navigation but the object definitions. In the Classic Console, it was much harder to view object definitions. Figure 5-19 shows how easy it is to view an object's definition details by clicking it.

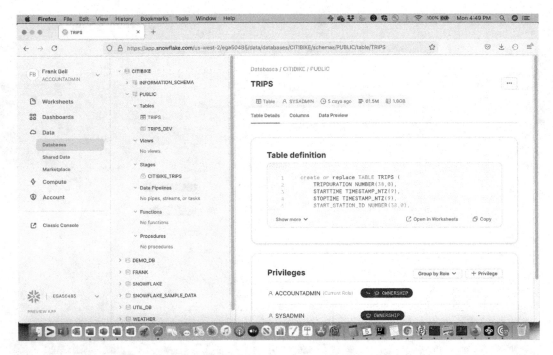

Figure 5-19. *Databases – Data Table Object Details Within Navigation Tree on Preview App (Snowsight)*

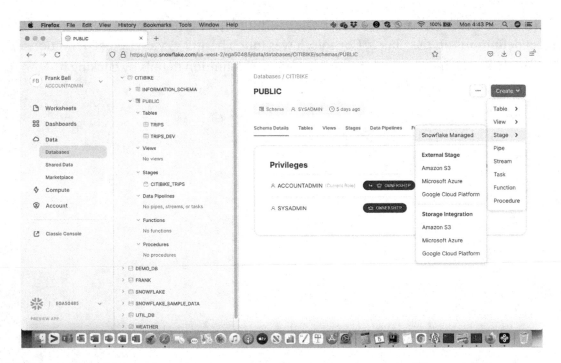

Figure 5-20. *Object Creation Functionality on Preview App (Snowsight)*

Shared Data

The Shared Data section under the Data navigation displays three subsections including "Shared With Me," "Shared By My Account," and a listing of "Reader Accounts." (This was the same Reader Account listing that used to be under Account but now has moved to a more appropriate location under Shared Data.) Figure 5-21 shows the interface with the Data ➤ Shared Data ➤ "Shared With Me" filter.

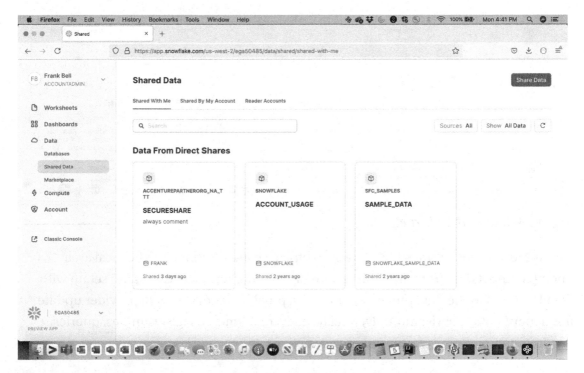

Figure 5-21. *Shared Data ➤ "Shared With Me" Example on Preview App (Snowsight)*

Data Marketplace

The Snowflake Data Marketplace is a fast-growing collection of open data and data available by request from various data providers on the Snowflake Data Cloud. Figure 5-22 displays the initial Data Marketplace visual view.

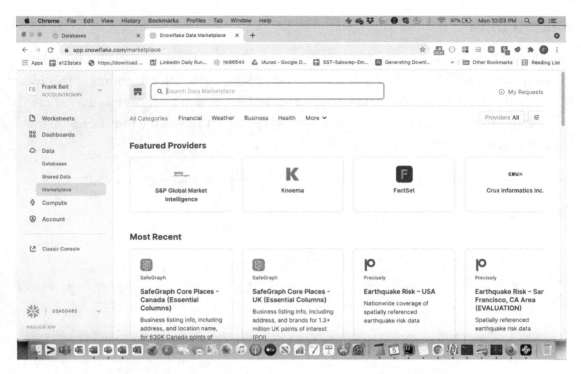

Figure 5-22. *Data Marketplace*

Figure 5-23 shows an actual Data Marketplace listing example of the local area unemployment data from ITS. This is how all Data Marketplace listings are set up with details related to the data provider name, data provider category, data provider update frequency (static, yearly, monthly, weekly, daily, real time), usage examples, queries, documentation, terms of use, and privacy policy.

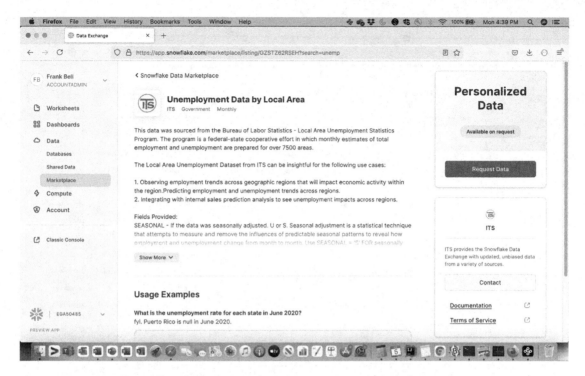

Figure 5-23. *Example of Data Marketplace Listing – Local Area Unemployment Data*

Compute

In Snowsight, Snowflake has created the Compute navigation section and placed Query History, Warehouses, and Resource Monitors within it.

Query History

History or now Query History has been moved under Compute in the Preview App (Snowsight). This is pretty much the same exact functionality as before but within Snowsight. Figure 5-24 shows the new Query History page in Snowsight.

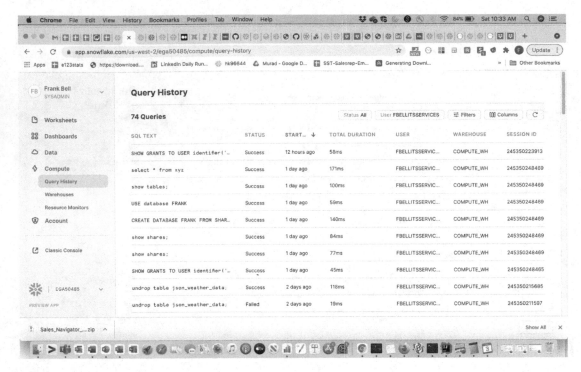

Figure 5-24. *Query History in Preview App (Snowsight)*

Warehouses

Warehouses (or really Virtual Warehouses) have been moved under the Compute
section. Figure 5-25 shows the new Warehouses page in the Preview App (Snowsight).

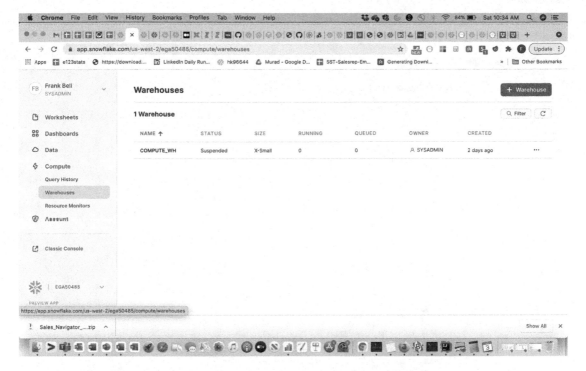

Figure 5-25. *Warehouse Listings in Preview App (Snowsight)*

Resource Monitors

Snowflake has moved Resource Monitors into the Compute section of the Preview App (Snowsight) instead of having it in the Account section where the Classic Console had it. Since this is such a critical component of managing costs, this is much better for Snowflake users. Figure 5-26 shows the new Resource Monitors page in Snowsight.

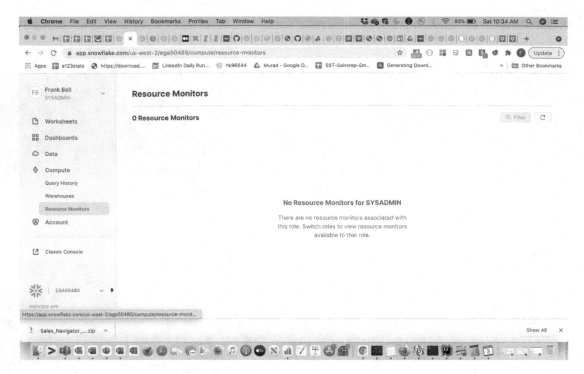

Figure 5-26. *Resource Monitor Listings in Preview App (Snowsight)*

Account

The Account view has become more transparent in the Preview App (Snowsight) and now shows what capabilities are available for users besides the ones using the Account Admin role. There are four subsections under the Account view now including Usage, Roles, Users, and Security. (You will notice that both Resource Monitors and Reader Accounts have been moved to more relevant navigation areas.)

Usage

The Snowflake Usage visuals have improved with Snowsight and made it easier to navigate and view your usage of compute, storage, and data transfer. Figure 5-27 shows compute usage, Figure 5-28 shows storage usage, and finally Figure 5-29 shows data transfer usage. You also have a lot more flexibility to view usage costs with date filters than you had in the Classic Console which had very limited Usage reporting.

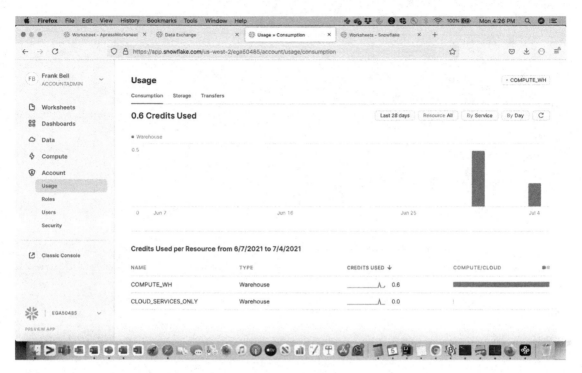

Figure 5-27. *Compute Usage in Preview App (Snowsight)*

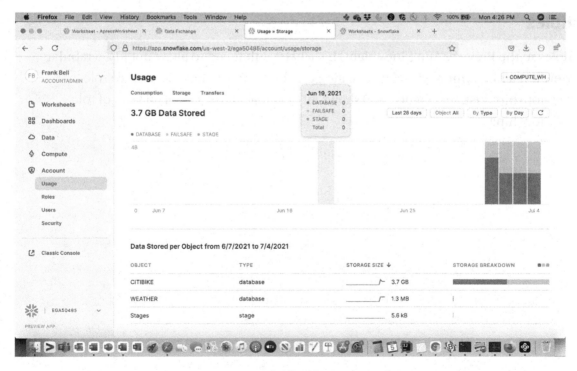

Figure 5-28. *Storage Usage in Preview App (Snowsight)*

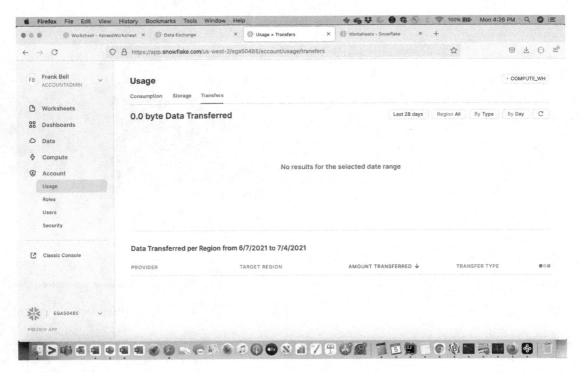

Figure 5-29. *Data Transfer Usage in Preview App (Snowsight)*

Roles

Roles in the Preview App (Snowsight) have been enhanced to allow not just a tabular view but a much more visually appealing graphical view to be able to trace role relationships visually. Figure 5-30 shows an example of the graphical view of role relationships.

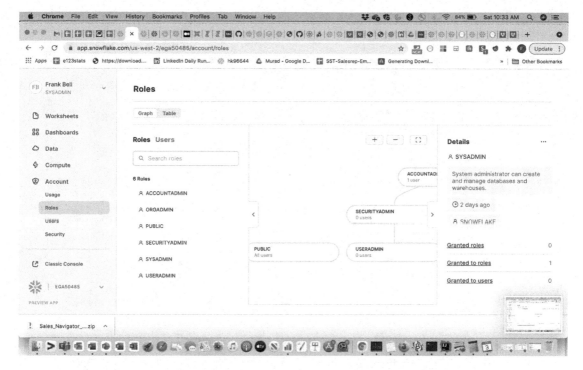

Figure 5-30. *Roles Graphical View in Preview App (Snowsight)*

Snowsight still kept the tabular view of roles as well in the interface. Figure 5-31 shows what the Preview App (Snowsight) displays for traditional role tabular view.

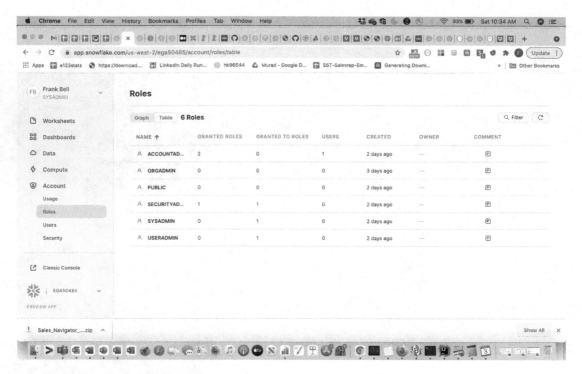

Figure 5-31. *Roles Tabular View in Preview App (Snowsight)*

Users

The Preview App (Snowsight) kept user listings pretty similar to the Classic Console. Figure 5-32 displays the user listings view in Snowsight.

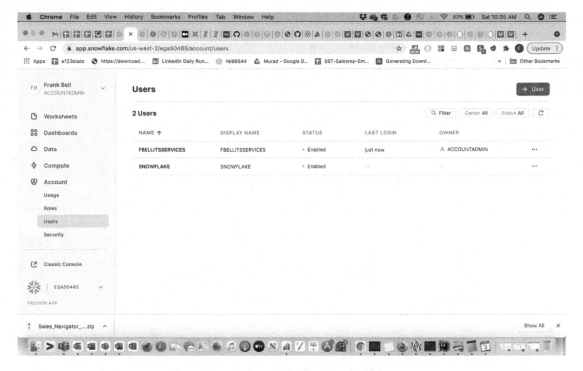

Figure 5-32. *User Listings in Preview App (Snowsight)*

Security

Network Policies (previously Policies in the Classic Console) and Sessions have been moved under a section named Security in the Preview App (Snowsight). Figure 5-33 shows the Network Policies screen within Snowsight. This is where you can add IP filters as network policies.

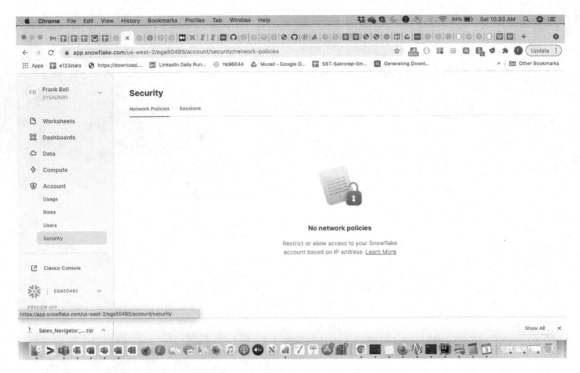

Figure 5-33. *Security View in Preview App (Snowsight)*

Snowsight Additional Navigation

You will also notice that there is a Snowflake icon on the bottom left of the Preview App (Snowsight) where you can use the dropdown to navigate to other accounts that you have access to. You still have to reauthenticate, but this makes it a lot easier to navigate to additional accounts you have access to in any region.

Snowsight also has more shortcuts. Figure 5-34 shows the shortcuts available for worksheets on Snowsight.

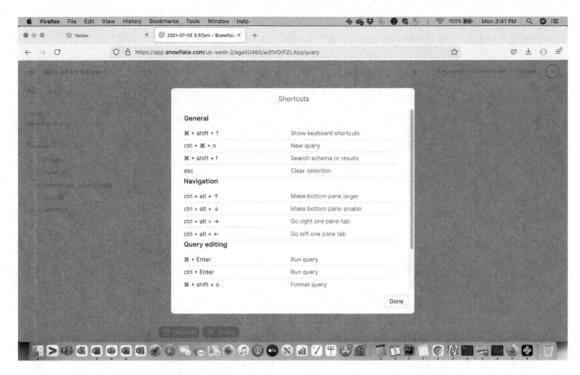

Figure 5-34. *Shortcuts for Worksheets*

Summary

Snowsight is the new Snowflake web interface and has many new and useful features
that are unavailable within the Snowflake Classic Console including the capability
to share both worksheets and dashboards. Another great feature of Snowsight is the
combination of both autosuggestion and autocompletion of SQL and Snowflake objects
within the worksheets. We made every effort to cover all the features available at the
time of writing, but we know this interface will continue to evolve to make querying and
administration of the Snowflake Data Cloud easier and easier.

CHAPTER 6

Account Management

If you have already created your trial Snowflake account or have access to it with the elevated ACCOUNTADMIN privileges, you will first need to understand how you can protect your Snowflake account, as well as your expenses.

As technology has progressed, so has the need for understanding and implementing good security practices, and while we will cover security in more detail later in this book, this will set the foundation as well as help you understand the available options and tools for managing and administering at the account level.

Despite being a relative newcomer in a crowded field of database technologies, there is a surprising set of tools that come standard with Snowflake that will help the Account Administrator manage their environment to secure and protect it.

In this chapter, we will cover the following topics:

- Traditional database administration

- The Snowflake Paradigm

- Risk mitigation

- Knobs, levers, and switches

- Security basics

- Monitoring your usage

So let's dig in.

Traditional Database Administration

Back in the good old days of saddles, spurs, ISAM, and indexed relational database management systems (RDBMSs), databases were highly technical and specialized tools that required a lot of knowledge and experience to get the most out of them.

© Frank Bell, Raj Chirumamilla, Bhaskar B. Joshi, Bjorn Lindstrom, Ruchi Soni, Sameer Videkar 2022
F. Bell et al., *Snowflake Essentials*, https://doi.org/10.1007/978-1-4842-7316-6_6

For example, here is a list of things traditional database administrators must know:

- Networking: Diagnosing network connectivity issues between users, database servers, application servers, authentication systems, and backup systems; opening ports in firewalls; checking latency; etc.

- How to securely authenticate users according to corporate security policies using the tools provided by the database platform

- How objects were going to be stored, including which physical devices would be mapped to logical devices and which specific RAID levels would be needed, depending on how critical the data was vs. the needed performance

- Which parameters needed to be set and what settings were appropriate

- How to secure and protect the database against hackers as well as human error

- Planning for business continuity and disaster recovery

- Enabling databases to "scale" by adding more storage and servers, which were dependent on what was available in the data center and, if adding new servers was going to happen, how long would it take to get those new servers installed and configured – as long as 6 months depending on hardware availability

Finding employees who had all of these skills was a difficult proposition, and when you did find them, they were very expensive to hire. And, if you accidentally hired someone who was weak in any area, it could mean disaster with extended downtimes, potential for loss of data, and poorly performing databases that did not adequately provide for growth, not to mention lots of very unhappy users.

In short, databases were hard. Very hard.

This also meant that highly experienced employees who had all the necessary skills to manage these databases were also expensive.

I was once asked this question in an interview for a position managing a very expensive traditional (not Snowflake) RDBMS platform: "If you could change anything about the platform and its technology, what would that be?"

I didn't even have to think through the answer, and I responded: "I would make it more complicated."

They looked at me incredulously and asked: "Why on earth would you want to make it more complicated?"

I again responded right away with: "Because I could make more money."

The moral of this story is that complicated systems require extensive knowledge to support, and therefore it becomes harder to find qualified individuals, as those salaries will be considerably higher to lure them away from their already lucrative positions.

So not only are traditional databases very hard to support, they are also very expensive to support, not only for the support contracts needed from the vendor but also for finding highly qualified personnel to help manage them.

The Snowflake Paradigm

The people who built the Snowflake database platform knew all too well the pain points of database administrators, having worked for one of the largest and most respected database vendors on the planet: Oracle.

They knew firsthand all the challenges listed previously and wanted to build something that was already secure but also easy to manage.

They understood that in order to be successful, the key was not in making databases complicated and hard, but rather making them easy to use and manage.

This meant that the traditional model would need to be thrown out and a new method of administering the database needed to be created. One of the ways they did this was leveraging the automation made available within cloud environments, through the various storage and compute cluster resources offered.

In the cloud, they could automate a lot of the tedious administration tasks to quickly and easily deploy and scale their offerings to enable companies to face their challenges and overcome them without the lengthy planning and hardware lead times and become much more agile in the process.

Risk Mitigation

If you will be administering and managing security within a Snowflake database, you need to be aware of the risks involved and how to avoid them.

As administrators, part of the job is protecting and securing the environment against

- Data loss

- Hardware failures

- Outside threats (e.g., hackers)

It's our job to do everything in our power to ensure that the data is always available, in our control, and out of the hands of those who are not allowed to get hold of it.

There are lots of tools provided by Snowflake to make it easier to lock things down, but first, let me share a list of potential risks:

- Users with insecure passwords

- Using the same password for both work and personal accounts

- Using the same password for both your bank and personal accounts, such as email

- Leaving network ports open, allowing foreign network access

- Oversharing on social media

- Giving out information without vetting the person requesting it

- Co-workers sharing passwords for the sake of convenience

- Sending people's private and confidential information via email using "Reply All"

- Clicking unverified links in emails, or the email addresses themselves

- Leaving your laptop, tablet or phone, or other device unlocked and unattended

- Leaving room for someone to see your screen and read over your shoulder

- Surfing non-work-related websites on a company-provided device

- Leaving your laptop visible in an unlocked vehicle with the keys on the dash

I could go on forever, but you get the idea.

If you can imagine a threat, there's a high probability that it can be prevented fairly easily.

The job of the administrator is to imagine the worst and then plan for it.

Knobs, Levers, and Switches

In the old movies portraying mad scientists, they would show them frantically turning knobs, throwing levers, and moving switches. This gave an impression of a series of complicated actions that would undoubtedly result in something bad happening, leading to some monster being created and unleashed on an unsuspecting public.

Those knobs, levers, and switches represented different aspects of control, all working together to produce a specific result.

Databases have their own sort of knobs, levers, and switches, and these are called *parameters*. Each database platform has a number of different parameters, each able to have two or more different values that can be used, each with a specific purpose.

The last I checked, Oracle has well over 300 parameters that can be adjusted to affect how an Oracle database works.

By comparison, Snowflake has – as of this writing – less than 70, with about one-third of them pertaining to date and time formatting options.

You can find a complete list of these parameters on Snowflake's website:

```
https://docs.snowflake.com/en/sql-reference/parameters.html
```

Parameters fall into three categories:

- Account Parameters: Global and high-level parameters that control account security, objects, and user sessions

- Object Parameters: Affecting databases, schemas, tables, and other objects' behavior

- Session Parameters: User-modifiable settings controlling formatting, collating, caching, and result management

If you have access to your Snowflake database, whether you have the ACCOUNTADMIN role or a lower role, you can see some, if not all, the parameters, using the following command:

```
SHOW PARAMETERS;
```

This will show each parameter, with its name, current value, and default value, and may also show a brief description of what that parameter controls.

Typically, parameters can be set at the account or object level and, in certain cases, can be overridden by a user session to allow different handling of the data from what might have been set at a higher level.

For now, we will focus on those parameters that affect the Snowflake account. Other types of parameters will be addressed in chapters dedicated to their specific needs.

Account parameters fall into the following categories:

- Security

- Data handling

- Date and time

While we'll be tackling the topic of security in more detail later on, for now we'll just take a look at a few important parameters to consider regarding security.

NETWORK_POLICY

This parameter is responsible for defining which IP addresses are allowed to connect to the Snowflake account.

While this particular parameter will not be very helpful for someone with a trial account who is connecting to Snowflake through their home Internet provider, where their IP address is changing dynamically, it is VERY important for companies where IP addresses are static – meaning they don't generally change.

The chapter that focuses on security will go into more detail, so for now, consider this parameter as being responsible for providing Snowflake with a built-in firewall for protecting the account.

PERIODIC_DATA_REKEYING

This parameter is much less complicated than NETWORK_POLICY in that it only accepts BOOLEAN (meaning true or false, 1 or 0, on or off, etc.) values.

The default is FALSE, but when set to TRUE, Snowflake will rotate the encryption keys annually and ensure that retired keys are destroyed once they are no longer needed.

While we will address encryption in more detail later on, we are generally in favor of anything that enhances the security of our systems and therefore recommend setting this value to TRUE.

This can be done using the following SQL:

```
ALTER ACCOUNT SET PERIODIC_DATA_REKEYING = TRUE;
```

It can be set back to FALSE at any time, but keep in mind that once encryption keys have been destroyed, they cannot be brought back and used again.

Since the keys are rotated on a regular basis, this should not be an issue since those old keys would no longer be applicable.

QUOTED_IDENTIFIERS_IGNORE_CASE

If you have ever worked with Unix, you may be acutely aware of how case sensitivity of text can impact files and processing, because while to humans the words Text and text may look the same, in some operating systems, the difference is significant.

Likewise, in some RDBMS platforms, case sensitivity can also impact the interpretation of object names.

A very common example is Microsoft SQL Server, where an object named "[Table].[Column]" is completely different from "[Table].[column]".

Except in SQL Server, you can refer to "TABLE.COLUMN" (note that we are using all uppercase AND have dropped the square brackets), which could be understood to mean either "[Table].[Column]" or "[Table].[column]".

Snowflake however is much more literal when it comes to case sensitivity.

If you specify that a column is named 'Object' within single quotes, Snowflake will take that to mean that the column must use that exact case for any future references, so that if you query the OBJECT column, Snowflake will generate an error saying that the column does not exist or that you don't have the necessary permissions to see it.

If your company typically references columns using a variety of case handling (e.g., first letter capitalized, all caps, or even "camel case," where one letter near the middle of the name is uppercase, e.g., "myObject"), then you might consider using the QUOTED_IDENTIFIERS_IGNORE_CASE parameter.

This parameter is Boolean too and tells Snowflake to basically relax and not take life so seriously.

The net effect is that Snowflake will not be so specific when it comes to quoted object names and allow referencing them in any style of case handling you choose.

The drawback however is that if you have columns that are similarly named, "Object" and "OBJECT", for example, it could be confusing and Snowflake may not know which object you are referring to.

To take advantage of this parameter, use the command

```
ALTER ACCOUNT
SET QUOTED_IDENTIFIERS_IGNORE_CASE = TRUE;
```

Note also that this can be set not only at the account level but also for individual sessions, where newly created objects will ignore case-sensitive references to names, as well as in queries.

However, if setting at the account level, it will only affect objects created from that point forward or until the parameter is subsequently disabled.

Unless your company is migrating from SQL Server and depends on this case sensitivity for its object identifiers in reports and other data processing, our recommendation is to set this parameter to a value of "TRUE".

Otherwise, you may want to set this value to "FALSE".

PREVENT_UNLOAD_TO_INLINE_URL, REQUIRE_STORAGE_INTEGRATION_ FOR_STAGE_CREATION, and REQUIRE_STORAGE_INTEGRATION_FOR_STAGE_ OPERATION

When it comes to loading and unloading data, there are a few ways this can be accomplished within the Snowflake environment.

While we will cover data loading and unloading later on, these particular parameters specifically address the security related to these types of data ingestion and storage management.

When data is loaded or unloaded to or from Snowflake, it is always sent to or pulled from staging areas in the cloud storage layer, which can be specified in a number of ways.

These three parameters specifically control whether or not users are allowed to load or unload data directly using S3 locations that are not using a storage definition known as a Storage Integration.

The ACCOUNTADMIN role is able to create Storage Integrations, secured with hidden authentication parameters, to provide a simple name for referencing these locations, rather than lengthy IDs and all the parameters that go with them.

These parameters force users to use these Storage Integrations and prevent them from using their own personal cloud storage locations, ensuring data is contained within the companies' digital borders at all times.

SAML_IDENTITY_PROVIDER and SSO_LOGIN_PAGE

These parameters are specific to Single Sign-On authentication, which allows users to leverage their corporate identity authentication infrastructure for allowing them access based on their network ID and authentication, whether it is a password or token-based system.

Typically, the ACCOUNTADMIN will work with the identity management team – the ones responsible for managing the infrastructure that provides authentication to all the systems in the company – for configuring this information.

These settings will be covered in more detail later on, but keep them in mind as we will come back to visit them when the time comes.

Security Administration

An important aspect – if not THE MOST important aspect – of account management is security.

The ACCOUNTADMIN role has the most responsibility in any Snowflake account, and it should not be any surprise that security falls directly into the lap of the Account Administrator.

There are a number of areas that the administrator should fully understand and take complete control over, and there are some that will only have minor tasks involved. They are

- Network protection

- Single Sign-On

- Data protection

- User login management

Of all of these, the least important – at least for the ACCOUNTADMIN – is user login management, which is really the responsibility of the SECURITYADMIN and USERADMIN roles, which will be discussed in greater detail in the next chapter on security.

For now, let's explore the first three options listed.

Network Protection

We touched on network protection in the previous section, with the NETWORK_POLICY parameter.

However, network protection is much more than just setting up a firewall.

Before we dig into those other aspects of this topic, let's look a little more closely at the NETWORK_POLICY parameter.

NETWORK_POLICY

As we mentioned previously, this parameter controls who can and cannot connect to the Snowflake account by explicitly defining IP addresses, much like a firewall does.

The IP addresses are expressed as a list of string literals, for example:

('10.1.0.1','10.1.0.2')

While the preceding example sets two IP addresses as being allowed to connect, it will usually not be useful to specify every single IP address. That would be tedious, particularly in larger organizations where lots of users and server clusters may need access.

Instead, it may be more useful to define a list of ranges, using CIDR (Classless Inter-Domain Routing) IP address range specifications, which allow specifying ranges of IP addresses using a short-form notation.

Unless you are on the network team for your company, you won't likely need to know the specific CIDR ranges that will be needed for your particular Snowflake deployment, but the network team will be able to provide the list of CIDR ranges and specific IP addresses on request using whatever communication tools they allow.

We will cover this topic in more detail in Chapter 7, when we discuss the topic of security in more detail.

Single Sign-On

You may have heard of this term before, as it is typically found in many corporate authentication systems.

It is a method of user verification that often relies on infrastructure using a combination of identity management tools to confirm that users are allowed to connect.

Snowflake is able to communicate with the more common methods of infrastructure authentication systems, such as Microsoft's Active Directory (also referred to as AD), Okta, Duo, and many others.

As long as the authentication system is SAML 2.0 compliant, the chances are very good that it can be used as the gateway for user access instead of passwords.

Because the ACCOUNTADMIN role is responsible for all security for their Snowflake account, it is their responsibility to work with corporate data security teams to configure Snowflake to work with their systems.

Data Protection

Another aspect of account management is doing what we can to protect the data from prying eyes.

Typically, database vendors would offer things like SSL connection options, encryption through expensive high-security packages, proprietary storage schemes, etc.

But with Snowflake, all data is encrypted all the time, whether in motion or at rest. In fact, it cannot store data in clear text at all.

However, there are certain things that an ACCOUNTADMIN can do that can be leveraged to enhance and strengthen the protection of the data stored in the database.

Here are a few things that can be done to help make the database as bulletproof as possible:

- Parameters

- Encryption

- Storage Integration

We covered a number of parameters in the preceding section pertaining to security, and it should be no secret by now that these are an easy way to help secure the account.

When it comes to encryption however, there *is* one area that Snowflake does not control, where encryption might not be configured: External (client-provided) Stages.

These stage areas are fully able to hold encrypted data, but the customer has the sole responsibility for protecting this data.

While it should already be protected by private encrypted keys, that may not be sufficient to please the data security team in your company, who may feel rather strongly about putting critical data "in the clear" (which means it can be read easily without any conversion) in cloud storage areas.

The way to keep them happy is to encrypt the data *before* it is written to cloud storage.

Encrypting data requires two things:

- The data to be encrypted

- A pair of encryption keys for scrambling the data

The "private key" is the master key, which is used to both encrypt and decrypt the data, while the "public key" is given to Snowflake (stored securely in Snowflake's encrypted system), which can decrypt the data.

When Snowflake pulls the encrypted files out of the cloud storage – still encrypted – it then holds them temporarily in memory, decrypts the data on the fly, and re-encrypts it with its own highly secure *composite* encryption key (basically several keys combined to make a superstrong key), before it writes the newly encrypted data to its own storage area.

And speaking of External Stages, ACCOUNTADMINs can create Storage Integrations that have authentication keys and cloud storage URLs associated with them, to make pulling data from external cloud storage accounts both easier and more secure.

The key to the Storage Integration is it simplifies the processes for loading and unloading data to and from cloud storage, because referring to a location using a simple name is much easier rather than having to include a bunch of parameters that are usually needed for both identifying and authenticating connections to a customer's cloud storage area.

When these parameters are used instead of a Storage Integration, whoever has this information could use it to load or unload data to or from that location, which is not considered to be a safe and secure practice. For example, having that information "in the clear," a user could email themselves that information and use it from their home computer to upload and/or download data, to use however they please – which tends to make data security teams cranky.

But when the location and credentials are given a name whose details are only known to the ACCOUNTADMIN, suddenly users no longer have unrestricted access to the location and can no longer copy the location and authentication information to use on their own, for their own purposes.

And to help further secure the database, the parameters we discussed in the section "Knobs, Levers, and Switches"

- PREVENT_UNLOAD_TO_INLINE_URL

- REQUIRE_STORAGE_INTEGRATION_FOR_STAGE_CREATION

- REQUIRE_STORAGE_INTEGRATION_FOR_STAGE_OPERATION

can all be used together to further strengthen and enforce the use of Storage Integration objects for loading and unloading data.

User Login Management

For most RDBMS platforms, databases exist to enable users to access data, to generate insights and understanding that is not necessarily clearly available.

We use SQL queries, reporting, dashboards, and analytics to pull information from the database, which means somebody must be consuming this information, which is often available as "self-serve," meaning users can log in to get access and run those reports or queries to gain the perspectives they need to help their company be competitive in the marketplace.

To enable those users to have access, we therefore need to have a way that we can give those users a unique and secure method of communicating with the Snowflake account. This is done through "logins."

Notice we don't call it an "account," as this might be confused with the overall Snowflake account that the company has purchased.

Logins provide a unique identity for users and can have attributes assigned to them to facilitate communication and permissions for accessing data and performing queries or other actions on data.

But it's important to realize that to be safe, we need to know how to reach a user in case of trouble.

For example, in a Snowflake account that I help manage and where I have the ACCOUNTADMIN role assigned to me, I discovered that a user had done something – probably not maliciously but rather by accident – that caused a warehouse to run constantly without automatically shutting down.

However, when the user's login was created, there were no additional attributes available to help identify who they were, and as such, we were unable to locate them.

We strongly recommend including the following attributes when creating new logins in your company's Snowflake account:

- First name

- Last name

- Email address

- Comment, which can include additional info such as phone number, corporate LAN (or network) ID, IM username, etc.

Now, when you first set up your Snowflake account, you will be given an ACCOUNTADMIN login by Snowflake that can be used for the initial access.

That login can be used not only for setting parameters, creating Storage Integrations, etc. but also for creating database logins.

Before any of those other items are done, it is strongly recommended that you create one or two logins for those people who will have the most responsibility for managing security and usage of the account.

Once they have been created, you can then grant them the ACCOUNTADMIN role, and those individuals should log in, immediately change their password from the default, *and* set up multi-factor authentication, also known as MFA.

Adding MFA to the login process adds an additional layer of security to the authentication process, which is important given the level of responsibility they will be carrying in regard to this database.

The strongest methods of security incorporate several factors that represent the individual in such a way as to make impersonating that user effectively impossible.

Separation of Duties

As was mentioned earlier, one of the things the ACCOUNTADMIN role allows is creating logins for other users of the Snowflake account.

The ACCOUNTADMIN has a lot of other responsibilities though, and when you have a lot of users having access to the database, invariably people can forget their passwords or lock themselves out of their login.

Enter the SECURITYADMIN role.

The only job of the SECURITYADMIN is to create other roles and logins and grant permissions to roles, as well as granting roles to logins.

There is also another role called the USERADMIN, which allows creating and managing logins.

It is strongly recommended that the only logins that the ACCOUNTADMIN creates are for the SECURITYADMIN and USERADMIN roles.

Once those logins have been created, then it will be the responsibility of the SECURITYADMIN and the USERADMIN to maintain the user logins and roles.

Only the role that creates a user login can change it.

This means that if the ACCOUNTADMIN tries to be helpful and creates a user login – perhaps the SECURITYADMIN and USERADMIN are out at lunch – then later when that user forgets their password, the SECURITYADMIN and USERADMIN will not be able to help the user regain their access, because the login was created by the ACCOUNTADMIN. Only now, the ACCOUNTADMIN is out to lunch or, worse, on vacation relaxing on a beach somewhere with no Internet access!

It is because of situations like this that we have something called "separation of duties." Now this applies to many other areas of IT, not just Snowflake login management, but it demonstrates that it's important to allow users with the appropriate responsibilities to do their jobs without interference or shortcutting defined processes and escalation trees. If the SECURITYADMIN and USERADMIN are out to lunch, the employee needing their access will have to wait until they get back.

Monitoring Your Usage

Another responsibility of the ACCOUNTADMIN is monitoring utilization within the Snowflake account.

This includes

- Storage consumption costs

- Warehouse credit consumption costs

- Replication and cloud compute consumption costs

- Ad hoc query monitoring

It costs money to run a business.

The larger the business, the more expensive it gets to hire people and set up the infrastructure that supports them, and there are costs involved for every part of that universe of systems and applications.

For example, a lot of companies love MySQL or MariaDB because it is open source and can be used without having to pay the high licensing costs of tools like Oracle and Teradata.

But it's not free, because the hardware – or cloud resources – still costs money to stand up, not to mention support licenses and the skilled personnel who know how to get the most out of it, how to best secure it, how to set up replication, etc.

Those tools are not optimized for the kind of scalability it takes to run a data warehouse like Snowflake, and piecing together all the components needed to make it work like Snowflake would be VERY expensive, not to mention the risk of implementing an untried system for production use.

Most database vendors provide tools – or ways – to track costs and expenses related to managing their platforms, and Snowflake is no exception.

There are two primary areas where money is spent with Snowflake accounts:

- Storage utilization

- Compute cluster utilization

Snowflake does not add on subscription fees, account management fees, user seat license fees, special option or feature package fees, etc.

Instead, based on the edition chosen, the organization that subscribes to Snowflake will pay based on the number of terabytes of storage used, as well as the number of compute node credits consumed, the latter of which depend on the size of the compute cluster chosen.

We won't go into the specific costs of compute clusters right now, but you can look forward to seeing this topic discussed in much more detail in the chapter that focuses on this subject.

However, it's only fair to be able to see and control – to a certain degree – how these expenses are incurred.

There are two tools provided that assist with this:

- ACCOUNTADMIN dashboard

- Resource monitors

The dashboard provides a view into the storage and compute cluster consumption of the organization, through graphical representations like pie charts and bar charts.

These charts can be filtered with date ranges so you can see where and when the bulk of the expenses have been occurring.

If you are currently logged into the ACCOUNTADMIN role, you can click the "Account" tab near the top of the Snowflake web UI page and then click the leftmost topic titled "Billing and Usage."

On this page are several areas where you can see what the consumption has been for warehouses, or you can change the view to see what the storage consumption has been for the various databases contained in the account.

By hovering over sections of the pie chart, for example, you can see the exact usage for the warehouses involved for the current period.

Likewise, you can filter these pages by date or see trends using the various perspectives offered.

Typically in Snowflake, database storage utilization expenses are relatively lower when compared with the compute cluster resources involved in maintaining the database.

Storage Costs

If you have already been monitoring your Snowflake account and know how much data has been loaded vs. how much is being stored in its compressed format, you can calculate the compression ratio and apply that to expected future growth to estimate future storage costs and utilization trends.

With this information, you can have an understanding of how much storage will be needed month to month and score discounts on future storage by committing to purchase that storage in advance.

In addition, you can also create your own custom queries to build your own perspective based on your organization's needs, using SQL and the new Snowsight dashboard system.

Now, that's all fine and good, but didn't we just say that compute clusters are more expensive than the storage component in the Snowflake architecture? Yes, that's correct. As a matter of fact we did.

So let's look at how we can manage those costs next.

Compute Cluster Costs

In any database platform, there is one area that is difficult to predict and protect against: the rogue SQL query.

This wouldn't necessarily have to be a big query in and of itself, but if it is written inefficiently, perhaps joining two huge tables on a calculated column, for example, this could potentially bring virtually any platform to its knees, and the database administrator's inbox would suddenly be filling with complaints and their phone would start ringing off the hook.

This can also affect Snowflake, but fortunately there are tools we can use to prevent this kind of thing from impacting other users, such as separating workloads to their own isolated warehouses, enabling automatic scale-out, etc.

But adding workloads and automatic scale-out are not going to keep those costs down... They will only minimize the number of calls the ACCOUNTADMIN receives, usually just for the person who wonders why their query is taking so long.

So what can we do to minimize our compute cluster consumption expenses? There are a couple of things we can do:

- Parameters

- Resource monitors

Parameters

There are parameters that can control the behavior of SQL queries and parameters that can control the behavior of the warehouses themselves.

These parameters are as follows:

- ABORT_DETACHED_QUERY

- STATEMENT_TIMEOUT_IN_SECONDS

- USE_CACHED_RESULT

ABORT_DETACHED_QUERY

This Boolean parameter controls how queries are handled if the connection to the login session is lost.

If the value is left as the default "FALSE", then queries will continue to run if the session connection is lost.

But if the value is changed to "TRUE", then if the connection is lost, the query is aborted after 5 minutes.

This is beneficial because it ensures that the system does not spend extra compute time on results that cannot be delivered to a user session that is no longer available, assuming data was being queried of course.

This parameter can be set at the account level or at the user session level.

STATEMENT_TIMEOUT_IN_SECONDS

This parameter uses the number of seconds provided to determine if the query is "running long" and needs to be stopped immediately.

The default value is 172,800 seconds, or two days, which may or may not be appropriate for the workload being processed.

This parameter can be applied at the account, warehouse, or user session level.

We recommend applying this parameter to warehouses, with a value that is reasonable for the work being performed.

For example, certain data loading processes may take a long time to get all the data loaded and in fact might be constantly loading data using a small warehouse 24 × 7, while an analyst working in an appropriately sized warehouse may not need more than 5 minutes to run their queries.

Therefore, it's important to test a broad variety of queries before setting this parameter, but we'll talk more about this topic in the chapter devoted to warehouses.

For now, if your Snowflake account is just getting up and running, we recommend starting with a smaller value, like 5 minutes, until you have a good sense of timing and scaling required for each of the workloads that the account will be handling.

You can always increase it later as needed, but remember that this parameter should be set to a number that represents seconds, not minutes or hours.

USE_CACHED_RESULT

This parameter controls whether or not the results of prior queries are reused for subsequent executions instead of being executed again, consuming credits by the chosen warehouse.

The default for this parameter is "TRUE" fortunately, which means that typically – unless changed – a query will return the same cached value that was returned the last time this query was ran, even if it was run by someone else, instead of consuming even more credits each time.

It can be set at the account, user, or session level and, if needed, could be changed to a value of "FALSE", which means that every time a query is run under this setting, a warehouse will be consuming credits and incurring charges.

Keep in mind too that just because this parameter defaults to "TRUE" does not ensure that it will use the cached results, because if the underlying data changed, it will cause Snowflake to recompute the results, incurring charges as a result.

Therefore, use this parameter with care, to make sure it is appropriate for the workloads it affects.

Resource Monitors

A resource monitor is a tool that allows the ACCOUNTADMIN role the ability to set thresholds for monitoring, along with predefined actions to be taken when each threshold is crossed.

You can see if there are any resource monitors defined using the command

```
SHOW RESOURCE MONITORS;
```

A resource monitor has one or more thresholds defined in terms of a budget of credits that the warehouse is allowed to consume, using a maximum credit value combined with percentages for each threshold.

For example, suppose you want to set a budget of 400 credits per month for a warehouse and you want to be alerted when it crosses the 50%, 85%, and 90% thresholds.

Before you can set those thresholds for a specific warehouse, you must first create the resource monitor to specify that budget and the named threshold percentages, like so:

```
CREATE RESOURCE MONITOR query_monitor
WITH CREDIT QUOTA = 400
FREQUENCY = MONTHLY TRIGGERS
ON 50 PERCENT DO NOTIFY
ON 85 PERCENT DO NOTIFY
ON 90 PERCENT DO NOTIFY;
```

The preceding SQL will allow an ACCOUNTADMIN to receive a notification whenever the QUERY_MONITOR resource monitor crossed a threshold.

So, when 200 credits have been used within a given month, a notification will be sent. Likewise, when 300 credits have been consumed, another alert goes out, and so on.

Now, if money is tight and budgets have to be adhered to, additional thresholds could be included to tell the warehouse to suspend operations or to even shut down immediately.

To do this, you could use the following SQL:

```
CREATE RESOURCE MONITOR query_monitor
WITH CREDIT QUOTA = 400
FREQUENCY = MONTHLY TRIGGERS
ON 50 PERCENT DO NOTIFY
ON 85 PERCENT DO NOTIFY
ON 90 PERCENT DO NOTIFY
ON 95 PERCENT DO SUSPEND
ON 99 PERCENT DO SUSPEND_IMMEDIATE;
```

You might have noticed we did not include thresholds for notification at 95% and 99%. This is because Snowflake will automatically send notifications on those types of actions.

Once a resource monitor has been defined, it can then be applied to a new warehouse, using

```
CREATE WAREHOUSE my_warehouse WAREHOUSE_SIZE = SMALL RESOURCE_MONITOR =
' query_monitor';
```

Or, to apply the resource monitor to an existing warehouse, use

```
ALTER WAREHOUSE my_warehouse
SET RESOURCE_MONITOR = 'query_monitor';
```

Now, suppose activity is high and the warehouse consumes 380 credits during the month, will this cause suspension of the warehouse since 95% of the budget has been consumed?

The warehouse will be suspended, and any new queries launched will wait in their respective queue until the resource monitor is changed to increase the budget allowed.

If any queries are in flight when the warehouse is suspended, they will be allowed to continue executing – and incurring charges – until they have either finished and returned their results OR until 396 credits have been consumed, which equates to the 99% threshold defined for the specified budget, at which time ALL queries running in that warehouse are aborted immediately and no further queries can be run until the budget is increased.

Keep in mind that the preceding examples are only to show different ways of using resource monitors. In fact, thresholds don't have to be limited to 100% but rather could be virtually ANY percentage that makes sense for your organization.

You might be wondering to yourself, why not set the budget to 100% before aborting?

This is because those defined percentages are not a hard limit, meaning that a few fractions of a credit could continue to be consumed while the warehouse is shutting down, and by setting the threshold to 99%, it gives us time to allow the warehouse to suspend before we hit the full budget we defined, particularly since this example was for a use case where our budget was tight and we wanted to make sure we stayed within the defined budget.

If this is a new deployment of Snowflake and you are just getting your feet wet, it might be worth setting up resource monitors with shorter frequencies, with perhaps a DAILY reset, so that if a budget for a particular day is hit, it can reset overnight and users can continue to work the next day.

While they wait, you can take some time to understand why the threshold was crossed, by investigating which users were using the impacted warehouse and diagnosing their queries to see if there might be inefficiencies in their structure that might have caused too many credits to be consumed.

It is recommended that at a minimum, one resource monitor be created for the Snowflake account and at least one for warehouses. This way, if a warehouse is missing a resource monitor, it can be caught by the account-level resource monitor, just in case.

Replication and Cloud Compute Costs

Although resource monitors are great for monitoring warehouses, there are other costs that can be incurred, specifically those called "cloud compute" resources.

These resources happen behind the scenes, for certain activities managed by the cloud services layer, where Snowflake has some additional work to be done, such as maintaining materialized views or managing replication between accounts.

Visibility into this type of consumption can be found in the Account tab, along with the warehouse Billing and Usage charts, but can also be queried from the ACCOUNT_ USAGE schema to get more detail on the underlying utilization.

Ad Hoc Query Monitoring

Finally, no discussion on monitoring would be complete without addressing how to discover what queries are doing up to the moment.

One of the tabs available in the Snowflake web UI is the "History" tab, which can display query activity along with some details like which role is running the query, which warehouse the query is running under, how long the query has been running, etc.

High-performing queries may only be a temporary blip in this window and quickly disappear with more queries taking their place, while other queries may take some time to finish.

If you have effectively separated your workloads to different warehouses to reduce queuing and workload impacts, you may still want to watch out for long-running queries or queries that have failed, such as timing out, thanks to the STATEMENT_TIMEOUT_IN_SECONDS parameter.

You can sort the results by various columns or even filter your results to eliminate queries that are of no interest, in order to narrow the list of results.

Using these methods, you can quickly see which queries are using specific warehouses and which are still running, and if you find one that concerns you, you can even click its query ID to find out more details about the query and how the query optimizer is executing it.

You may find opportunities for improving the performance of a query by rewriting it for efficiency or discover that the query is performing a full table scan because the columns being joined are non-deterministic.

Regardless, the History tab allows not only users but also the ACCOUNTADMIN to see what is actively and currently happening in the database to discover ad hoc opportunities for improvement.

Summary

Account management carries a lot of responsibility, not only for the security of the account and protection of the data but also for managing the expenses to a defined budget.

There are quite a few tools provided by Snowflake to make this easier, from parameters, Single Sign-On and MFA, network firewalls, and encryption to tools for monitoring resource consumption and expenses.

Always look to manage the account in such a way that security comes first, but there is more to the role of ACCOUNTADMIN than just security.

They also need to monitor the account as well as warehouses to ensure they are being used appropriately and effectively for separate workloads.

Finally, don't just monitor for credit consumption costs, but also find ways to keep track of cloud compute expenses, as well as user queries that might have opportunities for improvement.

CHAPTER 7

Security

Not everyone likes to talk about security... It's scary, intimidating, and so demanding that most of us would just rather not deal with it.

But if you think of it as something to avoid, think of it more like brushing your teeth, driving the speed limit, or doing your taxes. If you avoid it long enough, something bad is going to happen.

So it can be a strong motivator when we realize that allowing ourselves to neglect our digital environment by being careless with securing it could mean losing a lot more than what's in our bank account... It could take away your job, your family and friends...everything you love and have worked hard to achieve.

So let's embrace our inner warrior and carefully build defenses that will protect you and your company.

With that in mind, our goal in this chapter is to help you with those fortifications by covering the following topics:

- Security basics

- User creation

- User login authentication

- User roles

- Object permissions

- Securing network access

- Encryption

- Data protection

Before we delve into the mysteries of Snowflake security, it's important to discuss some basic rules for securing IT systems.

Buckle up, Buckaroo. We're jumping in, but not to worry, we'll start in the shallow end.

© Frank Bell, Raj Chirumamilla, Bhaskar B. Joshi, Bjorn Lindstrom, Ruchi Soni, Sameer Videkar 2022
F. Bell et al., *Snowflake Essentials*, https://doi.org/10.1007/978-1-4842-7316-6_7

Security Basics

Now, before we go further, let's talk about some very basic, but very worthwhile, security practices:

1. Firewalls

2. Keeping up-to-date

3. Antivirus

4. Phishing

5. Passwords

6. Emails with URLs

7. Unlocked devices

8. Using PINs

9. Protecting data

10. Backups

11. Encrypted communications

12. Social media

13. Social engineering

14. Monitor accounts

Let's examine these topics a little more closely, as you may not be familiar with some of them.

Firewalls

It is generally accepted that within corporate environments, firewalls are used to keep outside attackers or unfriendly organizations from gaining access to any systems and data.

Consumers, particularly those in the Information Technology sector, should always ensure that they have a router with a built-in firewall to protect themselves as well.

Even Snowflake has network policies that act as a type of firewall that can be used to tell Snowflake which connections are allowed to access an account or not.

We'll cover this topic in more detail in the section "Securing Network Access."

Keeping Up-to-Date

Let's be clear. I'm not talking about watching the news or listening to the radio and *definitely not* social media.

What I'm talking about is regularly checking your workstation, server(s), mobile devices, and personal computers to ensure they are regularly installing the latest system and package updates, bug fixes, and security patches, assuming of course that these are all configured to do this automatically, which is the generally recommended configuration.

If not, then they should be checked manually on a regular schedule, as this is how security holes get patched to prevent hackers from taking advantage of any back doors or other vulnerabilities.

When it comes to Snowflake, thankfully, there is no need to perform this task, as the company ensures that all systems receive prompt security fixes and patches, which are applied on a regular basis.

When a warehouse goes into a suspended state, the next time it starts, any new patches rolled out will automatically be in effect, even if the server was only suspended for a few seconds.

Antivirus

Another important layer of protection is antivirus, and like system updates, it's also important to ensure it is updated on a regular basis.

Its job is to look for anything that might get past the system and package updates and patches, such as an infected file being downloaded.

With regular updates, this helps protect all of these systems, both corporate and personal, from being misused by unauthorized access.

Fortunately, just as with being up-to-date on patches and fixes, Snowflake ensures that all their systems are protected with up-to-date antivirus software.

Phishing

This may be a new term to some of you, but no, this is not a misspelling.

Phishing refers to the act of trying to tempt a user into replying, clicking links, or opening files attached to messages – whether SMS, email, social media, etc. – in order to trick the user into revealing compromising information, such as usernames, confidential information, etc.

How big of a risk is this?

Back in 2020, officials at the highest levels of government were regularly tricked into replying to, clicking, and/or opening attachments in emails containing phishing messages, which resulted in one of the biggest hacks America had ever seen.

The hack, perpetrated by an unfriendly nation, impacted government – including the military – corporations, and infrastructure services (power, water, etc.) to such an extent that the damage is still being assessed.

Even though awareness was high as a result of the preceding occurrences, yet another government employee allowed themselves to become lax and opened yet another phishing email, resulting in yet again another hack, which fortunately did not do as much damage and was quickly mitigated.

Phishing is very real and has very real risks and consequences, and we should ALWAYS be suspicious of messages received in email or text messages or even via social media.

Passwords

We will go into this in more detail in the section "User Login Authentication," but here are a few basic rules to keep in mind:

- Longer passwords are better.

- Use mixed cases, numbers, and punctuation!

- Don't use dates like your anniversary, birthday, kids' birthdays, pets' birthdays, etc.

- Don't use words that relate to anything posted on social media.

- In fact, don't use dictionary words either.

The stronger the password, the harder it will be for someone to break into systems that you are responsible for.

Emails with URLs

Most of us have seen or even received emails from so-called "Nigerian princes" or government officials from foreign countries making too-good-to-be-true offers of riches to be deposited in our bank account, and I hope you have never fallen for these types of emails.

However, not all emails are that easy to spot.

They may appear to come from a co-worker or perhaps their personal email or an unfamiliar phone number or an online vendor like Amazon or Target or eBay.

There are a few ways to spot these, such as the following:

- Unfamiliar greetings that are out of place for the relationship

- Frequent misspelling of words or improper grammar

- Domain names in the From: address that look out of place or do not match who they say they are

- Domain names in URLs that don't match the company they claim to be

- A high level of urgency being expressed requesting quick action

- Attachments that are out of place or uncommon or not typical for the sender

- Requests for username, password, or other confidential information

When a sense of urgency is expressed, it is human nature to respond with compassion or fear and to quickly comply without taking the time to look for the warning signs shown previously.

By failing to recognize the warning signs, you put yourself and your company at risk.

Unlocked Devices

Leaving any device accessible and unlocked is highly risky.

How much do you trust strangers to leave your device alone and not browse through your pictures or social media?

For that matter, how much do you trust your co-workers?

I once worked for a company that had a couple of "jokesters" who thought nothing of hopping on a co-worker's unattended computer to bring up pornography sites in its browser because they thought it was funny.

ALWAYS lock your device if you are going to step away, but better yet, if it's a mobile device, then just take it with you, even if you are going to a "quick meeting" whether it's a phone or a laptop.

Using PINs

Similar to passwords, always secure your devices with a PIN that is more than just four digits whenever possible.

The longer it is and the more password-like it can be, the better.

Protect Data

When I was little, we had a cookie jar.

If I tried to sneak one behind my mother's back, she would call out "No cookies before dinner!" without even turning around.

It was like she had eyes in the back of her head.

Similarly, we have to be vigilant regarding who is behind us and what they can see on our screen.

If you are looking at pictures on your phone from your vacation, that's one thing; but if you have sensitive or confidential information on the screen belonging to a client or customer, then that information is at risk.

If, out of the corner of your eye, you see someone approaching whom you aren't sure you know and they don't continue on past but instead seem to slow down or linger, then close your app and politely ask if there is something they need that you can help them with.

Social Security numbers, usernames, addresses, and financial data are all considered information that should be protected at all costs.

Backups with Snowflake

Generally speaking, it is always good practice to back up your data, whether it is your work computer, personal computer, or data in a database you are working with.

You should never assume that data is being backed up. Instead, periodically check to make sure any recently added or changed files or data have been backed up.

Fortunately, with Snowflake, data is stored in the cloud in redundant locations, so that even if one data center is destroyed by a tornado, the data will still be available in another data center in the same region.

In addition, that data is also automatically backed up to a secondary storage area called "Fail Safe," which is kept for 7 days even if it is deleted.

Another win for Snowflake!

Encrypted Communications

In the old classic black-and-white spy movies, the hero sometimes might need to switch to a "secure line," where they used a briefcase that had a handset and another cradle that would convert their conversation into something that the enemy would not be able to understand or translate.

In computer speak, we call this encryption.

While we'll tackle this topic in more detail later on, this is a very important tool in protecting systems from being infiltrated, by using predefined "certificates" that each endpoint uses to shield the communications from others that might try to listen in.

Snowflake wins here as well, because all communications between users and application servers and Snowflake databases are always encrypted all the time and, in fact, cannot and will not accept any unencrypted connections.

Social Media

"What could pictures of kittens in teacups, dogs seeing a long-lost family member, or mean political memes have to do with security?" you ask.

Let me ask you this: have you ever posted about getting an interview – or better yet a job – with a company that you were excited about enough to say what its name was?

Guess what? Hackers can use that information to guess your role at the company and deduce what your username will be.

Combine that with all the posts of your pet that means everything to you, and they might figure out that your pet's name might be your password at your new job!

Social media is a feeding frenzy for hackers, and there is no low they won't stoop to in their quest to disrupt your job or even destroy your life, particularly if they think there is money in it for them.

Social Engineering

While almost everyone knows about social media, not so many are aware of the term "social engineering."

This technique often used by bolder hackers combines social media with direct contact with you or your loved ones.

For example, a young couple went on vacation and posted about it on social media.

Not long after, the parents were contacted by extortionists who knew the couples' names, where they were going, what they were going to be doing, etc. and informed the parents that they had been killed and that they needed to pay to have their remains shipped home.

The parents were devastated, but as they were about to send the money that was requested, the young couple arrived home from their vacation, to the shock and joy of their parents who fully believed the worst.

This story had a happy ending, but there are plenty of stories out there of people who have succumbed to scams like this.

Using direct contact combined with knowledge gained from social media posts, hackers and scammers will try to convince us or someone we know to divulge compromising information – not to mention the potential of financial risk – so they can use it for their own purposes.

It doesn't have to be someone being on vacation… It could be after business hours, in the form of an email pretending to be a co-worker asking for the files on the project you worked on containing financial details for a big presentation the next day.

The risks are high and the penalty is steep for ignoring the signs and being lax with both social media and phishing scams.

Monitor Accounts

Last of all, it's also important to keep an eye on your accounts, such as

- Social Media: Don't accept friend requests from people you don't really know, but ALSO beware of friend requests from people you "thought" you had already been friends with. Always check their profile to see if the account is new, because scammers will create dummy accounts that use your friends' profile information to try to trick you into accepting, which immediately gives them access to your friends' list, pictures, timeline, likes and dislikes, etc. Remember that topic we covered earlier about social engineering? So always keep your guard up.

- Online Bank Access: Check your online financial accounts regularly, and change your password once in a while too while you're at it. I once discovered a waiter in a restaurant I visited sent pictures of my credit card overseas, and suddenly I had hundreds of dollars of fraudulent charges appearing on my card.

- Online Shopping: Just like your online financial accounts, check your shopping accounts regularly as well. Everyone says their systems are secure until one day they appear in the news because they have been hacked, due to lax security precautions. It's also strongly recommended to not allow storing your credit card information to make checkout "easier." If they get hacked, it also makes it "easier" for the hackers to get to it as well.

Now that we've covered some basic best practices related to security, let's dig a little deeper into some of these, particularly as they relate to Snowflake.

Creating Users
Basic Login Administration

Creating simple user logins is fairly easy, if you have the appropriate role assigned.

This can be done via any tool that is able to connect to Snowflake and execute queries, whether using the Snowflake web UI, the SnowSQL CLI, or some other popular tool like DbVisualizer or Toad.

The key components of user login creation are

- Username and login name

- Identifying information

- Authentication method

- User default settings

Username and Login Name

The username and login name are used to identify the specific user within the Snowflake system. This information can be used to see when they logged in, which queries they have been running, which warehouses they have been using, etc.

They are similar but differ in that while the username is unique to the user within the account, the login name is associated with their corporate infrastructure identity and used in combination with a Single Sign-On system provided by your company's Infrastructure Security team.

When you log onto a corporate system, usually when you authenticate yourself to get access to various systems, you are communicating with some kind of Identity Provider (IdP).

Examples of Identity Providers are Active Directory (also known as AD), Okta, Duo, etc.

These systems ask for your password and sometimes may ask for a secondary method of identification, such as sending and/or requesting unique codes from a mobile device, also known as multi-factor authentication or MFA.

When Snowflake is configured with a SAML 2.0–compliant IdP, it no longer has to worry about handling authentication for those users, since the corporate infrastructure is responsible for that, as a centralized security service.

The login name is the user's identity that is associated with those corporate SSO systems.

Identifying Information

While technically optional, it is strongly encouraged to include one or (better yet) more information that helps identify users beyond a simple username or login name.

For corporate systems that may use more cryptic and less obvious usernames like a first initial + last initial + six-digit code in lieu of first name and last name, including identifying information can help find users in case of questions, such as when queries are running long, for example.

Identifying information includes first name, last name, full name, and email address, and the Comment field can also be used to store additional details such as user's physical office location or phone number.

Authentication Method

Specifying an authentication method is fairly simple.

If a user login will be authenticated by a password, you will specify an initial password, which they will be required to change on login (using the MUST_CHANGE_ PASSWORD = TRUE parameter).

If the login will be for an automated service account, then either it will be authenticated via OAuth + SSO, which doesn't require a parameter to be set during user creation – since this will be handled by the SSO system – or it might need the public key portion of the encrypted key pair to be specified, which Snowflake would then use for authenticating with incoming connections.

If SSO has been configured as an infrastructure service in your company, then the only thing that needs to match in that system is the login name specified to Snowflake.

User Default Settings

This information tells Snowflake which role, warehouse, database, and schema should be set as the user's "defaults" when they first log in.

If no defaults are defined, then they will need to specify them in their session before they can execute queries.

Without having default values set, users could become frustrated over having to constantly select their role, warehouse, database, and schema before running queries.

Therefore, it's considered best practice to specify these during user creation time, and if required, these can also be changed at a later time as needs change.

Snowflake User Creation

If you are using the Snowflake UI, you can do this from a worksheet using SQL or using the "Account" tab and then "Users."

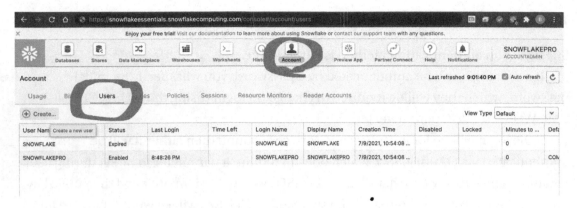

Figure 7-1. *"Account" and "Users" tabs*

Once you have navigated to this page, click the "Create ..." button on the left, to pop up a dialog box that will walk you through the creation of the user login.

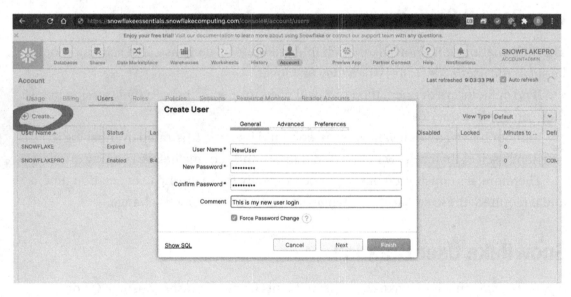

Figure 7-2. *Create User*

It's best to use a long password using a mix of numbers, upper- and lowercase letters, and punctuation.

Notice also in the preceding image, that the option for "Force Password Change" has been checked. The first time the new user logs in, they will be forced to re-enter the specified password, as well as to enter a new password.

Once created, the new user should keep their password private to themselves and not share it with anyone.

In particular, if a login for an automated service account is being created, then password authentication should never be used since passwords will periodically expire, and you don't want an automated production process dependent on a password that will expire. For these types of accounts, it's better to use either encrypted key pairs or OAuth authentication.

If you prefer to create users through SQL, where you have the opportunity to script the process, then this could be done using either a worksheet in the Snowflake UI or using their SnowSQL CLI.

Here is the minimum syntax required for creating logins in Snowflake:

```
CREATE USER username;
```

Note that there is no password or public key provided. For this user to be able to log into the Snowflake account, Snowflake would first have to have been configured for Single Sign-On with a corporate authentication service, and then the "testuser" login would have to be configured in that system as having permission to authenticate and access the Snowflake account.

In lieu of a corporate SSO service providing authentication for Snowflake, we can add either a password or public key as part of an encrypted key pair.

For this, we need to add one of the following parameters:

- PASSWORD = '<complex password goes here>'

- RSA_PUBLIC_KEY = '<public key string>'

If the user is going to authenticate using a password instead of SSO, then an initial password would need to be provided, for example:

```
CREATE USER testuser PASSWORD = 'cOmPl#x p4s$woRD gO3s hEr#';
```

Similarly, if the user will use an encrypted key pair to authenticate with Snowflake, then the public key portion would be specified like so:

```
CREATE USER testuser RSA_PUBLIC_KEY = 'MIIBIjANBgkqhkiG9w0BAQEFAAOCAQ8AM
IIBCgKCAQEAoLDhkm8wuCaA3gpe7QYWIuw8zwQJsNcTqkFGXnvrzcW+OODGhK3s5e5N0OA+
Jp8//FFHlKRp1l19bB2H/BfuCTOOuacJrRen60Jf2SCKy+qqws7aT8sl9DOP2f82ClKBNncawHM
1t4qdjvAwunz/GX2CBz515QoOpp43h0dVltr5YfUWMEHF6eMsGgOMUYBtR96jDMWYavRKgg1
J8vmjJ+CK2X1Sc3D+JAn7IRaYltHhXeHh9f3doM90Aph+kiRLwlY1w5SQI+7QNoxGaGoQffXf
vvrjLUfbmuau2RPkJUKGS81TKJr4SRdyfKPDQI7P+eWsMoqqhaYJFAROlKOaSwIDAQAB';
```

Note that the encrypted key must be a 2048-bit RSA public key portion in pkcs8 format and entered without the first and last lines, for example:

```
-----BEGIN PUBLIC KEY-----

... and ...

-----END PUBLIC KEY-----
```

When specifying the public key, make sure to exclude those lines, and put all characters on the same, uninterrupted line from the first to the last character, as shown in the preceding example.

Including other details such as identifying information and default settings is just a matter of including the appropriate parameters for each portion of the desired information.

See Snowflake's documentation for more details on creating users:

```
https://docs.snowflake.com/en/sql-reference/sql/create-user.html
```

User Ownership

Users are objects similar to other objects like tables, views, stored procedures, etc., and when they are created, the role that created them is their "owner."

With three roles capable of creating users and granting them permissions, it is strongly recommended that a consistent policy of role ownership be established, so that you don't end up with a random mix of owners of user logins in your database.

This can cause confusion when an urgent call for help comes in and the wrong role attempts – and fails – to reset a password or fix incorrect permissions.

To avoid this confusion, have a policy that outlines which roles are responsible for administering user logins such as the USERADMIN role and perhaps a separate role responsible for granting roles to user logins, such as SECURITYADMIN.

It is our opinion that ACCOUNTADMIN should only be used for managing the security at the account level and to allow USERADMIN and SECURITYADMIN to manage the day-to-day security tasks specific to users and roles.

User Login Authentication

Anyone who has touched a device that holds information will know what passwords are.

The ubiquitous password has become embedded in our lives, just like everyone has a vehicle of some sort. Even toddlers learn how to use vehicles whose motivation comes from their legs, then later graduating to bikes with training wheels, then to ones without training wheels, then to motorized vehicles, etc.

Likewise, some kids might get their first smartphone, and the second thing they should learn – after how to turn it on of course – are passwords.

If you have worked in the Information Technology field for any length of time or are even just paying attention to the news, then you are no doubt aware that hackers are constantly attacking government institutions and corporations, trying to find ways to break through firewalls, looking for back doors, or trying to compromise employees to trick them into accidentally giving them an opening into the corporate infrastructure.

A very large majority of corporations leverage, at a minimum, a user ID and password combination for identifying and authenticating users.

If a hacker is able to figure out an employee's username, perhaps from an email address – this was very common for a long time and very risky – then they have 50% of the information they need to get access.

From there, all they need to do is figure out the password, which they can often easily guess by using dictionary words or people or pets found in employee social media posts and profiles, for example.

If none of those work, then they are forced to start a method called "brute-force" hacking, where they begin trying similar words or names but including one or more character substitutions to try to guess a password.

There was a time when you could have a combination of a word, name, or number (e.g., could be a year, an age, or a date) and this would have been considered to be a pretty strong password.

But with the kind of computing power now available to the general public, this is no longer sufficient, and we need to use stronger methods.

Personally, I use an online password generator to automatically create a completely random string of 20–50 characters including upper- and lowercase letters, numbers, and punctuation, and then I selectively change some of the characters, numbers, and punctuation to further protect my new secret password.

Now, let's be real here for a moment.

You may have already thought to yourself, how are they able to memorize a 50-character password made up of completely random characters? Worse still, how would they be able to memorize more than one, for email, bank accounts, social media, online shopping, entertainment, etc.?

Honestly, I don't think 99% of the people could…and perhaps even 99.9999% of people couldn't unless they had a photographic memory.

What is needed is a secure place to store all of these passwords that is accessible to me whenever I need to look one up.

So, once I create a new secure password, I then take that password and store it securely in an encrypted tool called a password manager.

Some examples are LastPass, Keeper, NordPass, etc. but the idea is to use something that I can access whether I'm at home or out and about.

However, all of these may seem like a lot of bother for "just a password."

But there are a few more things we can use to enhance the security of our logins.

These factors should ideally include at least one of each of the following:

- Something You Know: This could be a password, PIN, passphrase, predefined questions and answers, etc.

- Something You Have: This might be a small keyfob that generates a series of numbers that is constantly changing, a cell phone using a special app for helping with authentication, a laptop, etc.

- Something You Are: This basically reflects something that is uniquely you, such as a fingerprint, retinal pattern, DNA, facial features, etc.

When all of these are used together, they form a highly secure method of identification.

For example, hackers may be able to figure out your password through various methods, or they may be able to glean enough about you to guess a passphrase based on your social media profiles, or they might be able to hack or hijack a cell phone, but getting a retinal scan or DNA sequence or fingerprint will be pretty difficult, especially if they are not even in the same country as you are.

Because of the level of responsibility that comes with having access to data – any kind of data – it's important to protect against those logins from the types of people who might be tempted to take advantage of the situation, if they should somehow be lucky enough to guess a password.

Therefore, we strongly recommend leveraging multi-factor authentication for more privileged accounts, including administrators, roles that own data structures, and roles that have access to personal or confidential data of any kind.

Now that we have discussed some basic password management rules, we should also address how to reset passwords for users.

Password Resets

If a user forgets their password, you can come back to the "Users" page to select their login ID and reset their password, again making sure the "Force Password Change" option has been selected.

Note that over time, a Snowflake login's password will expire and the user will need to change it before they can log in again.

In addition, users are not allowed to reuse a password within a certain period of time and must change it a number of times before they can reuse that password. However, reusing passwords at all is still discouraged as a best practice.

In most database platforms, such as Oracle, password enforcement is optional and can even be disabled if the administrator so chooses, but this is not optional in Snowflake, further solidifying Snowflake's position as the most secure database platform ever built.

Users can also change their password on their own, by going into their preferences.

Figure 7-3. *Preferences*

When the dialog pops up, they can select the "Change Password" option to change their password.

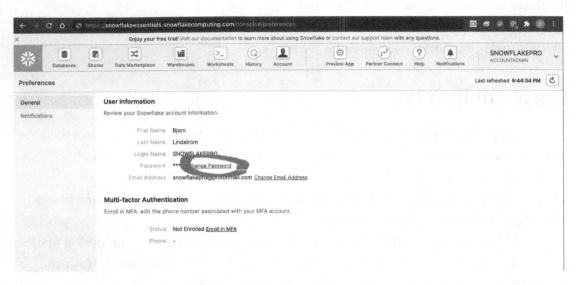

Figure 7-4. *"Change Password" Option*

Authentication Alternatives

Let's look at a hypothetical situation.

Suppose you have hundreds of employees that were created at the same time, using a scripted process.

Snowflake is laser focused on security, and as part of any comprehensive security plan, user passwords expire after a period of time.

Once that period of time has been reached, all those users will need to reset the passwords, and a lot of them will be reaching out for help with resetting their password.

But what if instead of having to help users reset their password, they could use the infrastructure that your company may already have in place for managing user passwords and controlling authentication with corporate systems, not to mention enabling automatic login capabilities?

This is known as "Single Sign-On" or SSO.

With this infrastructure in place, it allows authenticating with Snowflake with all the hard work handled by that platform rather than Snowflake.

When users' passwords expire, they only need to reach out to the corporate help desk and use those tools for their reset, and when Snowflake is properly configured with those SSO platforms, you no longer have to be the one holding users' hands for their password reset.

The topic of SSO is a big one and usually requires the help of other teams within corporate environments, and if you are like most people who are trying to learn, you probably have a trial Snowflake account and will not likely have the resources on hand for doing your own SSO configuration anyway.

Therefore, rather than doing a deep dive into this topic and how to configure Single Sign-On, we'll save that level of detail for another time.

Now having shared a ton of info on passwords and password management in the preceding sections, it's important to note that not all connections to databases are made by humans for working with data and processes.

Companies almost always have other automated processes that need to authenticate with databases to perform data loads, prepare dashboards and reports, etc.

These processes use different backend connection points that bypass the web GUI front end using interfaces like the SnowSQL CLI, ODBC (which stands for Open Database Connectivity), and JDBC (similar to ODBC but Java programming based).

These methods of authentication are predefined, often using configuration files that define all the pieces of information needed, such as the name of the database account, login username, password, role, database, warehouse, schema, etc.

This is a much more technical topic that really demands a lot more of a deep dive into the weeds than we have space for here, so we'll save that for another book.

Now passwords are really not a very secure method for automated processes, because of the ease of copying the password and using it to log in directly through the web UI.

Instead of passwords, it is better to use encrypted certificates, which are also known as "private/public key pair" certificates.

These can only be used through SnowSQL, ODBC, or JDBC and when properly locked down are a much more secure method for automated processes.

Another risk to automated processes is that passwords will periodically expire, which could be very disruptive, particularly when no one is tracking those password resets and it takes your ETL data pipeline or executive dashboard down unexpectedly.

Because of this, encrypted certificates are a much more attractive option.

User Roles

In order to secure objects within databases, it is a common practice with most database platforms to allow granting permissions to roles and then granting those roles to the users who are allowed that level of responsibility within the database.

These roles can usually be combined with other roles, and when granted to a user, all the permissions that were granted through all the roles are all active and available at the same time to the user.

Likewise, those platforms often also allow permissions to be granted directly to users, to accommodate special "one-off" or manager-level responsibility.

Users can also often own databases, schemas, and tables as well as other objects, and all permissions and roles granted are active all the time.

When it comes to Snowflake, it's not quite so simple.

But before I go into this topic in depth, I have another story to share.

One of my clients had an employee leave the company, and as part of the normal security processes, their login was removed along with all objects they owned in the database.

Soon after, important reports began failing; and as the investigation progressed to diagnose the outage, it was discovered that a critical table was missing.

It was not caused by any malicious intent, but rather carelessness in managing the ownership of the objects.

You see, the employee who left was accidentally set as the "owner" of the table, with all the necessary accesses granted to the production processes for accessing that data.

When their account was removed, all the data they were listed as the owner of were removed at the same time, which resulted in the outage.

Now, the security administrators should probably not have dropped all the objects by default... A process should have been triggered to find out what the objects were and what dependencies might exist for production processes, but it revealed a risk associated with users "owning" tables.

In Snowflake, users cannot own objects, with one exception: temporary session-restricted tables.

Instead, in Snowflake all objects are owned by roles.

This allows users to come and go, with no risk to the data if their account is disabled or removed completely.

Now you may ask, "But what if the role is deleted, either intentionally or by accident?"

Snowflake planned carefully for this potential as well, by ensuring that when a role is dropped, any objects that it owns have their ownership transferred to the role that was the owner of the role that was dropped.

If your head is swimming, not to worry. I'll explain it right here and now.

You see, roles are hierarchical.

This means that just like tables and functions, all roles are owned by another role.

For example, the ACCOUNTADMIN role owns the SECURITYADMIN role and the SYSADMIN role, whereas the USERADMIN role is owned by the SECURITYADMIN role.

The following diagram shows how these roles are related.

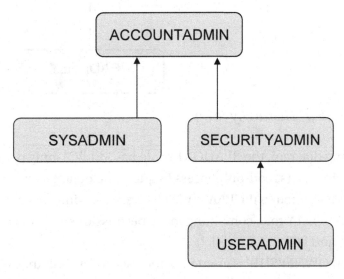

Figure 7-5. *User Role Hierarchy*

In the preceding diagram, you can see that USERADMIN belongs to SECURITYADMIN and likewise SECURITYADMIN belongs to ACCOUNTADMIN.

However, the "ownership" hierarchy is not the same as the "granted" hierarchy.

When you grant permissions to a role, the user that is granted that role automatically gets ALL permissions granted to the role, just as if they were granted to them directly, as long as they have selected that role in their session's context.

Likewise, if that role is granted to another role, then the second role also gets all of the permissions granted to the first role.

Consider this simple example.

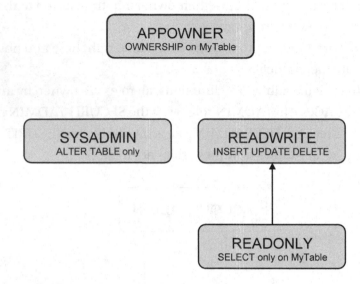

Figure 7-6. *"Granted" Hierarchy*

In the preceding diagram, the READONLY role has SELECT-only permissions granted to it, allowing users read-only access to query the data.

That role is then granted to the READWRITE role, which inherits the permissions granted to the READONLY role, combining those permissions with new ones allowing UPDATE, INSERT, and DELETE as well.

Note that when OWNERSHIP is granted to the user APPUSER, they can then do anything they please with the object, without the need to inherit the other privileges.

In fact, having OWNERSHIP allows the APPOWNER role the right to grant permissions to both the READONLY and READWRITE roles as needed.

Separately in that diagram we can also see that SYSADMIN has certain permissions to ALTER TABLE, but without the ability to even query the data.

This is a highly recommended method for ensuring separation of duties, where developers do not need to access data, but only manage the objects, such as adding columns.

Note that even better would be creating a separate role, perhaps named APPDEVELOPER that would have those specific permissions, but without the ability to create or manage warehouses as the SYSADMIN can do by default.

Object Permissions

Since we're on the topic of roles and granting object permissions to them, this would be a good time to briefly discuss the various types of objects and the actual permissions that can be granted to them.

As mentioned previously, all object access is granted to users via roles, and as such, no user can ever be granted a permission directly.

Instead, the necessary permission must first be granted to a role and then the role granted to the user.

If the user is logged in and using the role when the permission is granted, then the next time the user logs in or starts a new session and selects the role, then that permission will be available at that time.

There are three major areas where grants can be given:

- Account: This includes account-level objects like users, warehouses, databases etc.

- Database and Schema: Schemas are containers within databases that can hold objects.

- Object: This includes the typical table, sequence, stored procedures, File Formats, etc.

For accounts, typical permissions include USAGE, MONITOR, and MODIFY as well as the lesser used and more targeted OPERATE, CREATE SCHEMA, and IMPORTED PRIVILEGES.

Schema permissions include not only USAGE, MONITOR, and MODIFY but also CREATE for tables, sequences, stages, views, certain types of policies, and more.

Finally, object permissions include the traditional SELECT, INSERT, UPDATE, DELETE, and TRUNCATE, but also more obscure grants for stored procedures, tasks, pipes, streams, and stages.

Securing Network Access

We touched on network protection in the previous section, with the NETWORK_POLICY parameter.

However, network protection is much more than just setting up a firewall.

Before we dig into those other aspects of this topic, let's look a little more closely at the NETWORK_POLICY parameter.

NETWORK_POLICY

As we mentioned previously, this parameter controls who can and cannot connect to the Snowflake account by explicitly defining IP addresses, much like a firewall does.

The IP addresses are expressed as a list of string literals, for example:

('10.1.0.1', '10.1.0.2')

While the preceding example sets two IP addresses as being allowed to connect, it will usually not be useful to specify every single IP address. That would be tedious, particularly in larger organizations where lots of users and server clusters may need access.

Instead, it may be more useful to define a list of ranges, using CIDR range specifications.

You may remember from the Chapter 6 that CIDR stands for "Classless Inter-Domain Routing." Just to be clear, this acronym does not refer to an uncouth person wandering from town to town.

Rather, it simply allows referencing groups of IP addresses using a type of numeric shorthand.

For example, suppose we want to allow every IP address from 192.168.1.128 through 192.168.1.191. We could type every single one of those IP addresses out, wrap each one in single quote punctuation characters, and include them in our list.

Or we could instead use a CIDR range to refer to them, in a much more concise manner, like so: 192.168.1.128/26.

In a CIDR range, the value that follows the slash character indicates the number of bits that are used for network routing, when combined with the netmask to determine the specific IP address range.

Unless you are on the network team for your company, you won't likely need to know the specific CIDR ranges that will be needed for all the users and corporate systems that will interface with your particular Snowflake deployment, but that team will be able to provide the list of CIDR ranges and specific IP addresses on request.

If, however, you do find that you need to calculate ranges, you can find CIDR range calculators on the Web through your favorite web search tool.

Now that we have explored CIDR ranges briefly, let's go into more detail on the NETWORK_POLICY parameter, which has two options available, called ALLOWED_IP_LIST and BLOCKED_IP_LIST.

The list provided to the ALLOWED_IP_LIST option will specify – as you may have already guessed – the IPs allowed to connect to the Snowflake account.

Likewise, the list provided to the BLOCKED_IP_LIST tells Snowflake which IP addresses are NOT allowed to connect to the account.

It's important to note however that the BLOCKED_IP_LIST is processed first.

This means that if an IP address can be found in both allowed and blocked lists, then it will not be able to connect since the BLOCKED_IP_LIST rule is processed first, and Snowflake will not look for any exceptions resulting in "allowed" IP not being able to connect.

By specifying these lists, your database can effectively have its own built-in firewall to protect it against outside access.

The way this parameter is used is not by the ALTER ACCOUNT command, but through the CREATE NETWORK POLICY command, like so:

```
CREATE NETWORK POLICY MY_POLICY
ALLOWED_IP_LIST = ('ipaddr1','ipaddr2',...) BLOCKED_IP_LIST =
('ipaddr1','ipaddr2',...);
```

You can see the list of network policies using the SHOW NETWORK POLICIES SQL command.

It's important to remember to keep track of what each of these values or ranges cover, so that later on if something changes, you can easily identify what that range was for and where changes might need to be made.

For example, if the analytics team based in a remote city suddenly can't access Snowflake, it may mean that their IP addresses for their office may have changed and are no longer allowed by the network policy.

Having documentation to reference which resources are covered by the specified IP addresses and ranges will save you some headaches down the road.

Regardless if you like spreadsheets, text files, or online knowledgebases like OneNote or Confluence, make sure to track what is in the allowed and blocked IP lists so that if you aren't available, others have access to that information and can assist if needed.

Encryption

Back in the good old days of saddles, spurs, pencils, and paper, no one worried too much about passwords unless they were up to no good.

But in these modern times, not only do we worry about passwords, but we also worry about the data that crosses the globe in public networks, as well as when that data is stored anywhere a network can read, which is constantly being watched and searched by less than scrupulous people.

The way we hide information in that no man's land is through encryption.

This is one of the foundations of Snowflake's architecture and is part of every Snowflake account, free of charge.

Let me sum it up, right up front:

In Snowflake, all data is encrypted all the time, both at rest and in motion. Period.

Data in Motion

Let's start with data in motion.

This refers to information that traverses one or more networks, from one location to another.

When data is in motion but not encrypted, this is often referred to as "in the clear" or "clear text," meaning there is nothing preventing someone from easily reading those packets (individual groups of binary data).

However, when data is traveling in encrypted form, it cannot easily be decoded.

Note that I did not say that it "cannot be decrypted"... This is because there are some methods that people try to use as a form of protection that are really not safe anymore, such as using MD5.

However, there ARE secure methods, such as "SSL," which stands for "Secure Sockets Layer."

This is a very common – and popular – transmission protocol, and for good reason: it is very secure.

Just like when you make a purchase from an online store or check your bank account using your bank's website or app, these connections are made using SSL.

From the moment the two endpoints complete their "handshake" – the initial proof of identification between the two endpoints – all communications are encrypted, allowing complete privacy and security of data.

Regardless of whether the data is coming from Snowflake or traveling to it, no data moves until the connection is secured. Period.

Data at Rest

Another risk is where data is stored.

Most database platforms use their own proprietary File Formats, but that doesn't mean it is unreadable.

Legacy database files stored on disk or in the cloud can be read byte by byte to reveal data that may not have been meant to be seen without using the platform they were created by.

But if you encrypt the data and hide the key somewhere else – somewhere separate and secure – then it becomes virtually impossible to see the data.

In Snowflake there isn't just one key. There are several keys – a lot of keys actually – all stored securely and separately from data, as they should be to support their AES-256 encryption. This is known as a "composite key."

If someone managed to get their hands on all but one key, it still wouldn't be enough to read any of the data.

Every data file has its own key, which must be combined with a key that belongs to the table, which then must be combined with keys for the schema and database, as well as the account-level keys.

All of these keys combined form a key that must be used to decrypt the individual data file.

Regardless of whether the data is stored in the cloud services layer, warehouse layer, storage layer, or internal staging, all data is stored encrypted, all the time.

This has been mentioned before, but it bears repeating: This is a basic feature of Snowflake. Not an add-on package. Not an option that you can request as a "premium" subscription.

Free.

No charge.

Gratis.

Now THAT is a sweet deal.

But it doesn't stop there, because there IS an edition that offers even more security and control over the data: Business Critical edition.

This edition does cost a little more, but it adds one more encrypted key to the composite key used by Snowflake, to enable a feature called "Tri-Secret Secure."

Intended for organizations that hold critical client information, such as Personally Identifiable Information, or PII, health information, banking and finance, etc., this level of encryption requires not only all the keys used in Snowflake's composite key, but adds a separately owned and managed key provided by the customer.

If, for whatever reason, the client feels the need to shut down access to its data, such as a suspected breach of its network, the client can immediately disable their own key, therefore rendering all the data in the database unusable and inaccessible. Any attempts to query the database will result in an error indicating a problem with the customer's encryption key.

Once the customer decides all is well, they only need to re-enable their key, and all data becomes accessible and available again.

The process for enabling Tri-Secret Secure is not a trivial thing, like setting a parameter.

It requires several steps and working in conjunction with Snowflake Support to have it enabled.

But if your data is truly critical, then the Business Critical edition should be the minimum subscription level you choose when purchasing Snowflake.

Data Protection

Encrypted Stages

When loading data into Snowflake, the best practice is to use a storage area known as a "stage."

This is an interim place for holding data until it can be loaded into the database using COPY commands.

Although using Snowflake's built-in stages is easy and very secure, some companies prefer to use their own stages for this purpose, by leveraging bulk cloud storage such as AWS S3, Azure Blob, and Google Cloud Storage.

The danger in this is when companies fail to pay attention to securing these storage areas.

There have been many cases where private customer information has been stored without the proper security mechanisms in place to protect it and it has ended up for sale on the dark Web or, even worse, for free.

So how do we protect data in client stages? Using several methods of protection:

- Setting parameters to restrict stage definitions

- Securing the storage area

- Encrypting storage locations

The first item has already been discussed in earlier sections, so we won't spend much more time on that, but the next two are critical.

Securing the Storage Area

If you have worked as a developer, there are usually tight deadlines for accomplishing tasks, and there is a lot of temptation to take shortcuts to get things delivered, with the best of intentions to come back later and tighten things up.

Unfortunately, all too often there are new deadlines on the heels of the old ones, and we are back at the races again, and those good intentions most often fall by the wayside, in the rush to keep up with the never-ending onslaught of task after task after task.

Therefore, it's best to always take the time to do it the right way, straight out of the box.

So how do we do this with an External Stage?

There are some basic components to securing cloud storage based stages:

1. Location: Often a cryptic series of letters and numbers that form a unique name, identifying the location where files are to be stored.

2. Encrypted Key: This is a complex code that acts as a lock on the data, such that if you don't have the right key, you don't get access to the files, sometimes known as an SAS token.

3. Require encrypted HTTPS access only.

4. Always revoke public access.

Also be sure to check with your cloud provider to see what other options or features could be added to this to help secure your data even further.

Locking the data down such that Snowflake can't access it won't help things, so you don't want it TOO restrictive, but just enough so that only Snowflake and your ETL/ELT tools can access it.

Encrypting Storage Locations

Securing the location is a good idea but encrypting the location is even better.

This acts as a second layer of protection from unwanted access, and with Snowflake properly configured, it can be seamless and easy and protects the data by encrypting it when it is stored.

In order for Snowflake to access the data, it is configured with a set of keys, including a master key so that it can decrypt files read from that area.

This allows companies to secure their data, in their own storage area, to the same level of security that Snowflake uses within its own staging areas.

The process for setting up this encryption is different for each cloud provider, but the details for each can be found in Snowflake's documentation:

```
https://docs.snowflake.com/en/user-guide/security-encryption.html
```

Cloning

Another aspect of protecting data is ensuring production data is not affected by users, such as developers and testers.

In traditional databases, in order to isolate your production data from other impactful users, it is necessary to physically copy data from one place to another in your, yes, physical storage.

If it is only a few megabytes, or even a few gigabytes, that's not too cumbersome.

But Snowflake deals in terabytes of storage, which can take a long time to copy in any traditional platform.

But Snowflake has a secret weapon.

Its name? Metadata.

You might be thinking to yourself, "What the heck is that?"

You see, Snowflake stores a ton of information about each data file – also known as "metadata." And because the data files themselves cannot be changed, it's easier – and way faster – to copy just the metadata about each file to the new location so that the new owner has their own set of metadata while still pointing to the original files.

This allows the new owner to make whatever changes they see fit – which are written to completely new data files AND without touching the original, possibly production data files.

By copying only the metadata, there is zero storage penalty for the new owner, all while freeing the user to do with the data as they see fit.

I should point out, however, that the new "copy" of data is not completely free when it comes to changes.

If data is updated, inserted, or deleted or tables are modified, then Snowflake writes new copies of the affected micropartition files, and the amount of storage being used begins to grow.

The more the data changes, the more the storage utilization climbs.

More changes, more storage consumed.

But it's not just about shielding production data from accidental changes.

Cloning tables can also be used to create snapshot copies of the data, as an easily accessible form of backup.

Suppose you have a critical set of tables that you want to preserve at month's end – perhaps for regulatory purposes – that must be available for either reporting or recovery purposes for a period of time longer than time travel (another feature that allows for quick recoveries) allows.

You can create a clone of the table – or schema or even an entire database – for use later on.

This snapshot can be used for reporting – or recovering – as you see fit, without additional storage costs.

We'll dive deeper into this later on, but it's an amazing feature that is included at no extra cost with the Snowflake subscription, regardless of which edition you choose.

Data Masking and Row-Level Security

These final topics are specifically about protecting data stored in tables, more selectively than simply granting or revoking access to an entire table.

Dynamic Data Masking

Suppose you have a table that contains highly confidential data, which is protected by law and required to have security in place to keep it safe. We'll use the Social Security number as an example here.

When people call with questions about their account, they need to be able to prove they are who they say they are.

Let's say you call your bank. They may ask for things like date of birth, last four digits of your Social Security number, your phone number, or you address, all to confirm you are the actual owner of the account.

But have you ever considered whether they can see ALL your information? We may not actually know what they can and can't see, but we trust that their software systems protect us.

Snowflake has built a feature called Dynamic Data Masking, which enables this sort of protection to secure PII (Personally Identifiable Information) from being viewed when it is not allowed for the person querying it.

This is done through a policy that defines which roles have access to the column and what they are allowed to see within that column.

The policy can be defined such that the bank teller or the telephone banker may only see the last four of the Social Security number, so that when Snowflake receives their query, it knows what role has been assigned to the user and following the policy figures out that they are able to see only what is allowed according to the policy.

One role – perhaps the database administrator – may not be allowed to see any part of the SSN and be shown only blanks or perhaps "XXX-XX-XXXX" in place of the actual number.

Another role – let's say the bank teller – only gets to see the last four digits, for example, "XXX-XX-1234," which is the minimum amount of information they need to do their job.

Finally, the bank manager might be able to see the whole number, since they are in a position of authority and responsibility.

The policy described previously might look like this:

```
CREATE OR REPLACE MASKING POLICY ssn_mask
AS (ssn STRING)
RETURNS STRING ->
CASE
    WHEN CURRENT_ROLE() in ('BANK_MANAGER')
        THEN ssn
    WHEN CURRENT_ROLE() in ('BANK_TELLER')
```

```
      THEN 'XXX-XX-' || SUBSTR(ssn, 8, 4)
    ELSE 'XXX-XX-XXXX'
END;
```

If it is applied to a view for a specific column, then every time that column is queried from that view, the policy will make any changes to the data returned by that column to ensure it is protected according to the role currently in use.

When the policy is applied directly to the table however – which is the best practice – then regardless of whether stored procedures or views access the table, the policy will enforce the rules, even as data moves through the various layers, regardless of how it is accessed.

To apply a policy to a column in a table, a statement like this is issued:

```
ALTER TABLE user_info
MODIFY COLUMN ssn SET MASKING POLICY ssn_mask;
```

What we have described previously is known as column-level security, where we have secured a column according to the user's role.

Row Access Policies

But there is another way to protect data, and that is by controlling which ROWS are returned when a user queries a table, according to their role, which is known as row-level security.

In Snowflake, this is also known as Row Level Access, and is also controlled through a policy, called a Row Access Policy.

Similar to Dynamic Data Masking, the RAP is applied in context with the user's CURRENT_ROLE() that is in effect.

However, in addition to the role, a lookup table is used that correlates the rows of data to be protected, with the various Roles in use. If a Role should not have access to any data, then that Role would not have a corresponding entry in the lookup table.

Just like when you perform a JOIN between two tables, e.g.:

```
SELECT col1, col2
FROM organization O
JOIN departments D ON O.department = D.department;
```

Only those rows that have matching values are linked.

With row access policies, it is essentially adding another column in the JOIN, so that we add another control to further restrict the rows that are returned.

As an example, let's modify our preceding query to not only look at rows where we have a matching department value, but also restricting to those departments that our current role is allowed access to:

```
SELECT col1, col2
FROM organization O
JOIN departments D
ON O.department = D.department
   AND D.role_name = CURRENT_ROLE();
```

This is a very basic example, but as you can see, depending on what our role is, only those departments that have corresponding records matching our role will be included.

With row access policies, we create a policy that defines the lookup table relationship and apply it to either a view or table.

Similar to Dynamic Data Masking, the policy is first created and applied to either the view or the table. As before, applying the policy to the table itself is considered more secure and robust as a method of protection.

Unlike Dynamic Data Masking however, there are three parts required:

1) Table to be protected

2) Table with the column mapping

3) The RAP itself

For our example, let's continue with the preceding department setup.

Assuming we already have a table called "organization" that contains the "department" column, we want to restrict users so that they can only query their own department's organization info.

To do this, we will need another table that will be JOINed to the "organization" table, which we will call "departments," with the following definition:

```
CREATE TABLE departments
(department VARCHAR
,role_name VARCHAR);
```

Next, we'll populate that table with some departments and role names to allow filtering on some roles that presumably exist in our system:

```
INSERT INTO departments
(department, role_name)
VALUES
('HR','HR_ADMIN'),
('PRODUCTS','PRODUCT_ADMIN'),
('MARKETING','MKTG_ANALYTICS');
```

Now that we have our lookup table that matches the department names to our "organization" table, we can create a policy:

```
CREATE ROW ACCESS POLICY department_policy
AS (dept VARCHAR)
RETURNS BOOLEAN ->
'CEO_ROLE' = CURRENT_ROLE()
OR EXISTS (
          SELECT 1 FROM departments
          WHERE role_name = CURRENT_ROLE()
          AND department = dept
);
```

In the preceding example, if the CEO_ROLE is in use, then a value of TRUE is returned, since it satisfies the first condition.

Otherwise, the CURRENT_ROLE() is checked and compared to the department that was passed, and if there is a match on that column in the lookup table, then a TRUE condition is returned, and the current row is displayed or returned from the query.

To apply this policy to our organization table, we use the following SQL:

```
ALTER TABLE organization
ADD ROW ACCESS POLICY department_policy ON (department);
```

From this moment on, regardless if someone creates a VIEW on top of the "organization" table or uses a stored procedure or function against it, only the rows that are allowed according to the lookup table will be returned.

Summary

Now I need to be clear that the preceding examples of Dynamic Data Masking and row access policies are very simple and straightforward and there is a lot more to this than what I have described previously.

These mechanisms can get a lot more complicated – as complicated as anyone might need – but my goal here was to help you understand these features and the benefit they bring to enhancing the security of your database.

How you choose to implement it may differ in various ways, but always make sure to pay attention to access patterns and how users will be working with their data, and make sure there are no "loopholes" that they can take advantage of to get around the safety guardrails you have built.

CHAPTER 8

Database Objects

As a relational database management system, Snowflake supports a wide variety of objects, many of which you may have used in other database platforms, but not all.

These include the following:

- Warehouses
- Roles
- Databases
- Schemas
- Tables
- Constraints
- Clustered keys
- Views
- External tables
- User-defined functions
- Stored procedures
- Sequences
- Stages
- Ingestion pipes
- Data shares
- Tasks

Although this chapter will go into some detail on these topics, some of the topics deserve a lot more time and detail, and as such some will have entire chapters devoted to them.

© Frank Bell, Raj Chirumamilla, Bhaskar B. Joshi, Bjorn Lindstrom, Ruchi Soni, Sameer Videkar 2022
F. Bell et al., *Snowflake Essentials*, https://doi.org/10.1007/978-1-4842-7316-6_8

Warehouses

The engine of Snowflake – the part that actually executes the queries and loads data – is called a warehouse.

While Snowflake is a data warehouse, warehouses provide the compute power that is used for the heavy lifting in Snowflake.

These are basically one or more virtual servers provided by Snowflake to help execute any given query.

They come in a variety of sizes, but rather than describing them in great detail, such as CPUs, cores, memory, storage, and networking, warehouses are defined in terms of T-shirt sizes, with smaller sizes having less horsepower and larger sizes progressively doubling the amount of processing power available.

In addition, they can have one or several clusters of servers available to help manage query processing – also known as "nodes" – which can be automatically scaled in or out on demand as the warehouse fluctuates.

The administrators can grant or restrict user access to these resources as necessary, but generally speaking, you have to have a warehouse to query your data.

See Chapter 10 for a deeper dive into the mysteries of warehouses.

Roles

This topic was covered in much more detail in the previous chapter, but nonetheless they are considered objects just like warehouses.

Snowflake uses Role-Based Access Control, also known as RBAC.

They are very much like the roles used by other database platforms, but have some interesting differences that are intended to increase security and simplify maintenance and user access control.

To allow a user access to anything in Snowflake, that access – whether to be able to query a table or execute their query using a warehouse – must first be granted to a role. Once the necessary privileges have been granted, the role itself is then granted to the user.

It's been said before, but worth repeating here, that you cannot grant permissions on any object to any user. At all.

The only way a user can gain access to an object is for the access to be granted to a role and then to grant the role to the user.

This forces a much more planned and intentional security architecture that ensures that for any given role, users cannot bypass any built-in security controls by mixing permissions with multiple roles, since users can only leverage one role at a time.

However, you CAN grant a role to ANOTHER role and through inheritance allow users to cross role boundaries, but as mentioned before, this requires careful planning to ensure that only the permissions actually needed are available through the chosen role.

Out of the box, Snowflake provides a few roles that have limited capabilities, in part to ensure separation of duties.

These are:

- ACCOUNTADMIN

- SECURITYADMIN

- USERADMIN

- SYSADMIN

- PUBLIC

At the "top" of the food chain is the ACCOUNTADMIN role.

Note the use of double quotes around the word "top." This is because although ACCOUNTADMIN can do a lot of things, it cannot do anything/everything.

It can do a LOT, but if it has not been granted query privileges on a table, then ACCOUNTADMIN won't have access to that table.

This intentionally breaks from the tradition of the "God-level" role, because Snowflake takes security seriously, and users that are administering the database RARELY NEED to see, access or use sensitive data. After all, their job is to protect that data, but their job does not require SEEING Social Security numbers, for example.

Because the ACCOUNTADMIN role has full access to managing security within the database, it should be tightly controlled, as an individual with access to this could put a company at risk by making uninformed changes to the system.

Databases

Just like most other RDBMS platforms, databases are a high-level container for data, stored in tables and organized in schemas, like Figure 8-1.

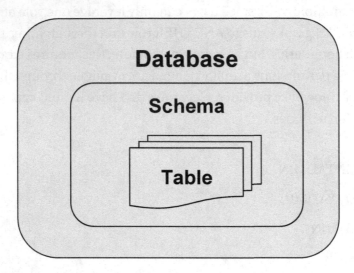

Figure 8-1. *Database, Schema, and Table Structure*

However, unlike most other platforms, Snowflake does not place any limits on the number of databases you can have in your account.

There are two kinds of databases:

- PERMANENT

- TRANSIENT

Permanent databases are just like they sound, intended to hold data long term. Conversely transient databases are short term and have limits on their recoverability.

For example, a PERMANENT table can have up to 90 days of time travel, plus another 7 days of Fail Safe available in case point-in-time recovery is needed, such as in the case of human error.

Transient databases on the other hand do not enjoy those features, having only 1 day of time travel available and NO Fail Safe.

And, because there are no other real benefits to using transient databases, they are recommended only for ELT purposes.

It is considered best practice to organize data into multiple databases and schemas – to the extent that it makes sense – to make it easier to maintain and manage objects.

Another way Snowflake distances itself from other vendors is that all databases are replicated across multiple availability zones automatically, to protect against data loss, so that if an AZ goes down, for whatever reason, your data is still available and does not require any manual failover or recovery.

If you desire, you COULD create a snapshot of your database by easily creating a clone, which will be covered in depth in another chapter.

Database clones could also be used for development purposes when production data needs to be protected.

For a truly robust disaster recovery architecture however, Snowflake's replication feature is very easy to set up and configure to duplicate databases not only between regions but also between different cloud providers, anywhere in the world that Snowflake supports.

Schemas

Similar to databases, schemas are another kind of container, which can hold most kinds of objects, and are themselves stored in databases.

Whenever a database is created, Snowflake automatically creates two schemas inside the database:

- INFORMATION_SCHEMA

- PUBLIC

The INFORMATION_SCHEMA contains views and functions that can be queried to find out information about the objects stored in that schema, also known as the metadata.

The PUBLIC schema does not contain anything, but can be used if desired to store objects.

In practice however, we rarely see this schema being used. More typically, specific schemas are created when required, with strict controls granted to roles as needed.

Similar to databases, schemas can also be transient in nature, with the same recoverability limits in place, and any objects created inside will also be transient in nature, which particularly applies to tables and the data they contain.

An interesting difference from databases is that Snowflake has another type of schema, known as managed schemas.

Typically, the role that has OWNERSHIP of objects stored in a schema is able to grant permissions on those objects to any role they choose.

However, with a managed schema, the role that has OWNERSHIP on the schema manages all privilege grants to other roles – which can include future grants – on all objects stored in the schema.

So, while the object can still be owned by a different role, that role cannot grant any privileges on the object to another role and instead must rely on the owner of the managed schema to grant any permissions on the object.

In addition, similar to databases, schemas can also be cloned when separate copies are needed for development or snapshot purposes.

Tables

Again, just like you find in other relational database platforms, Snowflake stores data in tables.

Depending on the platform, some vendors store their data in rows in the block and sector level on disk, but while Snowflake supports similar column data types as you would find in other systems, Snowflake stores its data in micropartition files, each of which contains data stored in compressed columnar format, along with metadata for the data stored there.

Once created in Snowflake's cloud storage, these micropartition files cannot be changed.

If a row stored in a micropartition table is updated or deleted, then a new micropartition file is created with the current data – along with a new date and timestamp and revised metadata – which becomes the newest version of the data.

Assuming the table is not specified as a TRANSIENT table, these timestamped versions are retained according to the retention period parameters defined at the ACCOUNT, DATABASE, SCHEMA, or TABLE level, and within that defined retention period, data can be queried as of that point in time if needed.

Tables can be created as one of three types:

- PERMANENT

- TRANSIENT

- TEMPORARY

Like databases and schemas, permanent tables can have longer retention periods than transient tables, up to 90 days if you have the Business Critical edition.

When it comes to temporary tables however, these are very different animals.

While both permanent and transient tables will persist until they are explicitly dropped by the owner, a temporary table will only exist for as long as the user session exists.

Once the user disconnects or the session ends, the table and all its data disappear.

There is no time travel nor Fail Safe available – more on these topics in another chapter – but their temporary nature makes them perfect for ELT/ETL processes where transformations or merging with other tables is needed.

Data Types

Snowflake supports a wide variety of data types for columns, but since they are ANSI standard, some conversion may be needed when migrating from other platforms.

Supported data types include numeric, string, logical, date and time, geospatial, and semi-structured.

Table 8-1 is a chart showing the various data types aligned with the categories listed previously.

Table 8-1. *Data Types*

Category	Data Type	Description
Numeric	NUMBER	Default precision and scale are (38,0).
	DECIMAL, NUMERIC	Synonymous with NUMBER.
	INT, INTEGER, BIGINT, and SMALLINT	Synonymous with NUMBER except precision and scale cannot be specified, with scale always equal to zero.
	FLOAT, FLOAT4, FLOAT8	Snowflake displays these database types as FLOAT even though they are stored as DOUBLE.
	DOUBLE, REAL, and DOUBLE PRECISION	Synonymous with FLOAT.

(continued)

Table 8-1. (*continued*)

Category	Data Type	Description
String	VARCHAR	Default (and maximum) length is 16,777,216 bytes.
	CHAR, CHARACTER	Synonymous with VARCHAR but length equals one character.
	STRING, TEXT	Synonymous with VARCHAR.
	BINARY	Capable of holding binary data, up to 16,777,216 bytes.
	VARBINARY	Synonymous with BINARY.
Logical	BOOLEAN	Currently only for Snowflake accounts created after January 25, 2016.
Date and time	DATE	Most formats accepted, but any time elements will be truncated.
	DATETIME	Synonymous with TIMESTAMP_NTZ.
	TIME	Stores hours, minutes, and seconds, as well as optional fractional seconds with precision up to nine digits for nanoseconds.
	TIMESTAMP	Synonymous with TIMESTAMP_NTZ.
	TIMESTAMP_LTZ	TIMESTAMP with local time zone, where any specified time zone is not stored.
	TIMESTAMP_NTZ	TIMESTAMP with no time zone, where any specified time zone is not stored.
	TIMESTAMP_TZ	TIMESTAMP with time zone stored.
Semi-Structured	VARIANT	Holds JSON, Avro, ORC, Parquet, and XML.
	OBJECT	Used to store key-value pairs, where key is a nonempty value.
	ARRAY	Used to store arrays of unspecified size, with non-negative indexes and values of type VARIANT.
Geospatial	GEOGRAPHY	Supports WKT, WKB, Extended WKT and WKB, as well as IETF GeoJSON.

Note that some of the data types are merely synonyms for the base data type.

With all data types, there is a hard limit of 16,777,216 bytes (16 MB), which is not necessarily the same as the number of characters, particularly when dealing with multi-byte character sets that may range from 4,194,304 (4 bytes per character) to 8,388,608 (2 bytes per character).

You may have noticed in the preceding chart a special data type called VARIANT, which can hold semi-structured data.

The 16 MB limit applies here also, which may require some special handling when parsing large JSON datasets, depending on their structure. You can learn more about working with semi-structured data in another chapter in this book specifically on this topic.

Constraints

Snowflake allows defining various kinds of constraints, including

- Primary keys

- Foreign keys

- Unique constraints

- Not Null constraints

Unlike many RDBMSs however, Snowflake does not enforce most of these, with NOT NULL being the only exception.

Note that I said they aren't ENFORCED... You can create them, but they will be more of a comment to provide users with information on typical access and JOIN patterns.

As such, there is nothing to prevent insertion of duplicate records into tables that, in their original form, may have had unique constraints in place intended to be a safeguard against this.

While this might be useful in a highly transactional database system, this creates undue overhead in a massively scalable environment that would have a negative impact on performance.

As an analytics processing environment, this platform is not intended to provide this kind of data processing and is more geared toward highly performant database queries against massive amounts of data, rather than focusing on one or two records out of trillions of potential records.

If this is a concern, it can be built into the ingestion framework that is responsible for loading the data into the system, but this is not the intended use of the platform.

Because some reporting tools and applications – and ingestion frameworks as well in certain cases – rely on constraints to provide hints as to the best method of accessing data within a table, Snowflake allows these constraints to be defined, even if they aren't going to be enforced.

All that being said, Snowflake DOES enforce Not Null constraints, which are, in fact, the only constraints enforced and honored.

Clustered Keys

In the same way that primary keys and unique constraints are not enforced in Snowflake, indexing in general is not even allowed.

Between their metadata and micropartition files, Snowflake does an excellent job filtering data (using the WHERE clause) as well as JOINing to other tables.

As such, Snowflake is generally faster without indexes and primary keys than most platforms that rely and even require them.

Not having indexes means also not needing hints in queries to suggest access patterns within queries.

However, there are some cases where access patterns may not fit the way Snowflake's optimizer thinks is optimal, such as edge case queries that are less common but still considered critical. Or perhaps over time the structure of the table or the data it contains shifts enough that the original optimization that Snowflake had chosen may no longer be appropriate.

For these conditions, Snowflake allows defining clustered keys.

A clustered key is a kind of metadata associated with a table that tells Snowflake which columns are ideal for accessing the data within the table or for JOINing to other tables.

It's not a primary key or index in the traditional sense in that it does not store an ordered set of keys as a way of looking up specific records.

Instead, as an analytics platform, Snowflake's focus is on quickly accessing huge amounts of data, so the solution is simply to name the columns in the order the data should be accessed, and Snowflake automatically handles the rest, with all subsequent queries attempting to use the new clustered key whenever possible.

Now, perhaps your table and its data have not changed over time and the original optimizer access patterns are still highly efficient, but there are a few queries that need to use different and non-optimal access patterns for that table.

It might be that creating a clone of the table might be sufficient, followed by a clustered key on the clone, or perhaps using a materialized view that has a clustered key might also work.

It's important to note that most of the time, clustered keys are not needed and they aren't even a best practice.

If you have tables that are multiple terabytes of data and higher (up to around 100–200 TB), clustered keys may be helpful, but only if you find your queries are not performing well, even after tuning the queries for efficiency according to the Query Profiler.

However, Snowflake does a great job optimizing queries, even for extremely large tables, and because of this there are very few use cases for implementing a clustered key on a table.

Views

Views are also of two types: normal views and materialized views.

Normal views are basically a named query that can be used to return columns quickly and easily, rather than having to repeatedly include the query in various forms.

In this way, you can build a common perspective of data in one or more JOINed tables that can more easily be leveraged than would otherwise be possible.

By giving a common query a name, users need only query the name of the view to return some or all of the columns made available in the view.

Materialized views share some common traits with normal views, except that there are data storage considerations – as well as certain limitations – in their use.

When a materialized view is defined, Snowflake actually stores the rows and columns that the materialized view would create in a separate set of micropartition files, just as if it was a table.

If your "base table" (the underlying table named in the materialized view) has a lot of columns and performing certain aggregations upon it, you could end up with a much smaller dataset that could be read and processed in a fraction of the time that the original table might have needed, helping users get their answers much more quickly as a result.

Unlike a view however, there are a number of restrictions that apply to materialized views:

- They can only query a table and cannot query a view of any kind.

- They cannot JOIN to other tables.

- They can only perform certain aggregations and grouping of rows.

- They cannot query a user-defined table function (more on this later in this chapter).

- They cannot include user-defined functions (whether internal or external).

- They cannot use Windowing functions.

- They cannot use the HAVING clause.

- They cannot use the ORDER BY clause.

- They cannot use the LIMIT clause.

- They cannot use GROUP BY on columns not listed in the SELECT clause, meaning ALL columns not being aggregated MUST BE IN THE SELECT LIST.

- They cannot use GROUP BY GROUPING SETS.

- They cannot use GROUP BY ROLLUP.

- They cannot use GROUP BY CUBE.

- They cannot use nested subqueries.

Regarding limitations related to aggregate views mentioned previously, the following aggregations are ALLOWED:

- APPROX_COUNT_DISTINCT (HLL)

- AVG (except in PIVOT)

- BITAND_AGG

- BITOR_AGG

- BITXOR_AGG

- COUNT

- MIN

- MAX

- STDDEV

- STDDEV_POP

- STDDEV_SAMP

- SUM

- VARIANCE (VARIANCE_SAMP or VAR_SAMP)

- VARIANCE_POP (VAR_POP)

Any other aggregation is NOT ALLOWED in materialized views.

Another variation of views and materialized views are known as secure views (or secure materialized views).

These types of queries protect sensitive information by hiding the underlying definition of the view, so that although columns and data are returned by querying the view (assuming they have the proper privileges granted to them), they would not be able to see which database, schema, or table (or view) the data is coming from.

Only users with the appropriate GRANTs – such as the OWNER – will be able to see the definition.

Stored Procedures and User-Defined Functions

There are two kinds of objects that can be created to perform special processing of data:

- User-defined functions (UDFs)

- Stored procedures (SPs)

A UDF is basically used for taking a parameter value that is passed in, such as a column name, and transforming that information in some way before returning it in a specific data type. Examples of a user-defined function might be calculating the radius of a circle or converting text from all uppercase to "Init Caps" where the first letter of each word is capitalized and the rest converted to lowercase.

UDFs can be written in either JavaScript or SQL and must return a value.

A stored procedure on the other hand is much more powerful, supporting complex logical operations on result sets generated from queries, and must be (as of the date of this book) written in JavaScript only. However, while they do not have to return a value, they also cannot be called from within a DDL or DML query, but rather must be called as a single standalone statement from the UI or the command line.

The good news though is that highly complex processing can be accomplished within stored procedures, including dynamic SQL, ELT and ETL data loading from stages, etc.

Most platforms have their own proprietary procedural languages (e.g., Oracle's PL/SQL and SQL Server's T-SQL), but Snowflake chose JavaScript because of its broad acceptance by the industry.

Snowflake's road map includes support for other languages down the road, but for now, JavaScript provides a powerful tool for complex operations, which include full Snowflake query abilities along with support for looping through those results built in.

Pretty cool, don't you agree?

There ARE some limitations however:

- External JavaScript packages cannot be imported.

- JavaScript does not support Parallel Threading – it is Single Threaded only.

There are other languages that will be supported down the road, but for now, this should keep us plenty busy for the time being.

Sequences

A sequence is an object with a single purpose: generating unique numbers across user sessions and SQL statements.

For example, suppose you are ingesting IOT data from the cloud that has only a device ID and a metric key and value but no event ID.

You could use a sequence to generate an incremental event ID as a default value on a column, which is automatically assigned when the row is inserted.

When used on large INTEGER columns (like using the BIGINT data type), this provides a broad set of values for historical trending purposes.

However, there are a couple of limitations that should be mentioned.

Because a sequence might be asked to provide a value from multiple concurrent sessions, a particular session may not get truly sequential values, resulting in gaps in the numbering.

Once a sequence reaches its maximum possible value, it will go back to the minimum possible value and begin incrementing from there again.

It is up to the developer to make sure to accommodate this possibility and provide solutions for occasional maintenance as needed.

Stages

A Stage is a type of storage object normally used during data loading and ETL processes.

While you could string a bunch of INSERT statements together to load data using an external tool, this is not recommended as it will not perform nearly as well as "bulk loading" from a stage, where massive amounts of compressed data can be loaded in parallel into the database at once (more on this later in the book as we've dedicated an entire chapter to this topic).

Stages can be either internal (managed by Snowflake) or external (managed by the customer).

Internal Stages are always encrypted by default and cannot be used unencrypted, which ensures that no matter what kind of data is stored there – even long term – it will remain safe and protected.

You can create an Internal Stage and give it a name, or you could create a Temporary Stage that disappears at the end of the session.

In addition, stages can be privately owned by a user or be specific to a table or be used to hold incoming (or outgoing) data for a number of databases, schemas, and tables.

On the other hand, External Stages may or may not be encrypted, depending on user requirements, and are managed by the customer within their cloud provider's services.

Because of this, securing and managing those storage locations is the sole responsibility of the customer, since Snowflake has no ownership of those locations.

Each cloud provider offers at least one and possibly more storage services that can be leveraged as External Stages:

- Amazon AWS S3

- Microsoft Azure Blob

- Google Cloud Storage

The safest way to use these stages is to create a Storage Integration, where the credentials are stored encrypted in Snowflake's metadata and cannot be queried or reverse engineered.

Just like tables, stages are created within schemas and can be referenced using a similar notation, like Database.Schema.Stage.

One of the things that makes both permanent Internal and External Stages so convenient is the ability to specify the formatting of the data files they contain, so that it doesn't have to be specified directly during the COPY statement.

File Formats

A File Format is used to specify the structure of the data files that contain the data to be loaded to (or unloaded from) the stage.

You can use a predefined standard format type, including

- CSV

- JSON

- AVRO

- ORC

- PARQUET

- XML

Or, in some cases, you can apply or alter these definitions to meet the requirements of the incoming data, such as accommodating text that is wrapped in double quotes but also might contain embedded double quotes.

There are some parameters that are available that can tell Snowflake how to interpret the incoming data files.

We can set a wide variety of parameters to set attributes, such as

- Compression type

- Field and record delimiters

- Whether headers should be skipped

- Date, time, and time zone formatting

- Encoding

- Null handling

- Special cases such as columns exceeding the width of a column in the target table and so on

File Formats can be defined with a specific and descriptive name to make referencing it easier during loads, by specifying just the name of the format instead of listing all the parameters for every COPY.

Pipes

Unlike so many of the objects listed previously that are used in various ways to store or reference stored data, pipes are objects that enable automatic loading of data into tables without having to schedule jobs (whether from tasks or an external data ingestion framework), as the files become available in the queue.

This allows getting the data into the database without having to issue COPY commands periodically.

For example, if you schedule a COPY command to run once an hour, you may or may not have data to be loaded, but the command still runs and consumes compute time in a warehouse in order to perform that check.

But if you have defined a pipe to watch a specific storage location and only execute a COPY command when data arrives, you could ensure data is loaded into the database within minutes of its arrival, without the need for constant "wake up and check" processing.

Like other objects, pipes are assigned a name, but have a COPY statement associated with them that specifies where the incoming data is located, as well as the target table.

Pipes are able to leverage the cloud provider's built-in messaging services so that as new files arrive at the location where they will be picked up, Snowflake will be notified that there are files waiting for processing and will proceed to execute the associated COPY statement to load the data.

These data files are then loaded into a queue where they are processed in a continuous and server-less (meaning without needing a warehouse) manner according to the parameters defined for the pipe.

External Tables

An external table differs from a normal table found in a schema in that the data is sitting in a stage or other Storage Integration somewhere, just like compressed files that have been copied to a stage in preparation for loading into Snowflake.

In fact, if you have files that you have already ingested OR files that have YET to be ingested, you can use them as the basis for an external table.

You can define the columns used in the external table based on the File Format chosen, and once created, you can query it just as if it were any other table including JOINs to other tables or creating VIEWs on top of them.

Using the various parameters available, you can limit which files are included and, for huge amounts of data, may be able to enjoy higher-performing queries as a result.

As an example, we just set up an external table for a client, whose queries in Snowflake were already running in a fraction of the time – a few minutes – that they needed in their legacy system, but when we recreated the table in full with smaller Parquet formatted data files (using Snappy compression format), we were able to reduce the query time to seconds.

Snowflake uses the original files located in a cloud storage data lake (S3, Blob, or GCS) and uses its metadata to tell the optimizer about their structure, similar to the way it does for normal tables, including the file location and path, version identifier, and partitioning information.

Because of this, Snowflake can query these files as if they were a part of the database itself.

Now, you may be wondering to yourself, "If external tables are so fast, why would I want or need to ingest the data into the database at all?"

The reason for this is that external tables are READ ONLY. You cannot make changes to them. No UPDATE, no DELETE...no changes. At all.

There may be a need for querying original, raw source data, particularly when analyzing the evolution of data from legacy systems into the next-generation data warehouse. External tables are a perfect tool for this type of analysis.

Now, the results I described previously were not necessarily typical and only represent a recent experience with external tables, and your results may vary depending on the data, format, compression, row organization, etc.

Remember that Snowflake stores its data in highly efficient compressed columnar format, and data provided from legacy systems is rarely going to be in such an optimal state as we had in our testing, so remember that external tables are considered to be less efficient and less performant than Snowflake's proprietary micropartition files.

Although we have never tried this, let alone tested it, if your performance is not what you would like it to be, you might try creating a materialized view on top of the external table and see if that might work.

Tasks

Unlike many of the objects described previously, tasks are a way to schedule either SQL or stored procedures, but they can ONLY run a SINGLE STATEMENT when they are triggered.

Fortunately, you can also create tasks that are dependent on the successful execution of another task. These are called the "root" or "parent" task, and their dependent tasks are referred to as "child" tasks.

A parent task can have any number of child tasks, but child tasks will only execute if the parent task finished successfully, without registering an error.

By creating a hierarchy of these dependencies, you can string together any number of tasks in order to fully define the needed workflow.

Through careful planning, you can create very simple workflows using tasks, such as ingesting data from a stage into a table.

With even more careful planning and well-thought-out requirements, design, and JavaScript coding, you can create very complex workflows that can do validation on incoming data, ingestion into transient tables, flattening of semi-structured data into views and/or tables, complex transformations, and logging of progress and results, if you are ambitious enough.

Tasks run on a predefined schedule, using a predefined warehouse and using the role that created the task in the first place.

Not only will that role need appropriate permissions on the objects it will be working with, but it will also need specific privileges that will allow it to create the tasks in the first place, which are

- EXECUTE TASK on ACCOUNT

- USAGE on DATABASE

- USAGE, CREATE TASK on SCHEMA

- USAGE on WAREHOUSE

It is considered best practice to create a specific role that will include the preceding privileges on the database(s), schema(s), and warehouse(s) that can then also be granted a role that includes any other necessary permissions needed for accessing the rest of the objects needed.

One interesting aspect of tasks is that they do not run as the user that created them.

To prevent situations where a production workflow stops working when a user is removed from the database, all tasks run as the SYSTEM user, with the role that created the tasks controlling the access.

When looking at QUERY_HISTORY (as ACCOUNTADMIN if no other roles have been granted the necessary permissions), you will see the SYSTEM user executing the SQL or stored procedures using the role that created them, on the warehouse that was specified in the tasks' DDL.

There are a few ways to specify when a task should be run.

Parent tasks can use simple timing notation, such as "30 minutes" or "600 minutes."

For more complex schedules that may need to run on different days of the week, for example, you could use a CRON-style scheduling syntax, in the form of

```
<MINUTE> <HOUR> <DAY-OF-THE-MONTH> <MONTH> <YEAR> <DAY-OF-THE-WEEK>
```

This format is very flexible, allowing multiple variations of each of the parameters shown, like using the asterisk character ("*") to indicate that every one of the event-type parameters it is representing should be included in the schedule.

For example, if a schedule is defined as "0 0 * * * MON,TUE,WED,THU,FRI", then that means that the task will run at midnight (hour zero and minute zero) of every day of every month of every year when the day is Monday through Friday. If the day is Saturday or Sunday, then the task will never run.

You can do a Google search on the term "CRON examples" to see how to create more complex examples for more information.

When it comes to Daylight Savings Time, this represents a unique challenge, since it can cause unintended effects, such as a job getting run more times than needed in the fall, when the clock is turned back an hour, or jobs getting missed and skipped completely, in the case of having a job that runs at 1:30 AM on a Sunday morning in the spring, when the clock jumps forward an hour.

If your job is critical and MUST run only once and every weekend at 1:30 AM, then you will likely need to do some careful planning around the Daylight Savings Time changes, to ensure your critical job runs every time and only runs once.

Now, suppose you have scheduled your ELT task to run every 30 minutes, but at some point, someone dumps a massive amount of data into the stage you are using, and your task runs for hours processing it all.

What happens when the 30-minute mark comes up?

Normally, Snowflake will ignore that event (regardless of whether the task is the parent or one of the dependent child tasks that are still running) and wait until all the tasks have finished their work.

Rather than immediately kicking off another task to try to "catch up" followed by more tasks trying to do the same, Snowflake will pretend that those events were not necessary and will launch the task when the next scheduled event comes up.

This is called overlap protection, which can be managed to a certain degree to allow specific instances of overlap to occur, but the normal behavior is to ignore any missed events in favor of the next scheduled event.

Also be sure to pay attention to your warehouse sizing, because if there are other users – or if you have multiple child tasks planned to run at the same time – that are going to be using the same Warehouse, you may need to do some testing to find the optimal size.

The best practice is to isolate workloads to prevent queuing, particularly if there are SLAs in place for getting processing finished.

Summary

Snowflake provides a broad array of objects that bring all the functionality provided by most database platform vendors, and in some cases even more.

With support for database, schema, stage, table and column level storage combined with other objects like sequences, stored procedures, tasks, pipes etc. you will find all the necessary components for building data warehouses that can meet even the most stringent requirements.

Combining these with appropriate security controls such as hierarchical role based access control, secured views, and data masking and row access policies, it can be every bit as secure as any company might need.

CHAPTER 9

Querying and Cloning Data in Snowflake

In this chapter, we will demonstrate how to query and clone data in the Snowflake Data Cloud. We will provide examples of SQL queries to show you how to get required information out of the data in tables and views within Snowflake. You will also learn how to clone data within Snowflake, which is one of its key differentiated features.

The Snowflake Worksheets in the Snowflake web interface is where we will demonstrate querying and cloning data. In Chapter 4 we went over the Classic Console Worksheets, and in Chapter 5 we went over the Snowsight Worksheets. You can execute Snowflake SQL queries from the web interface, SnowSQL (Snowflake command-line interface), third-party BI tools, and any connector.

Snowflake follows the American National Standards Institute (ANSI) Structured Query Language (SQL) standard. If you are already used to the standard ANSI SQL queries, then it should be a simple transition to Snowflake for you to query and analyze data. In addition, Snowflake has its own set of functions that massively help to analyze data faster. We find that data professionals who have used other relational database management systems (RDBMSs) before find it very easy to transition to query and analyze data within the Snowflake Data Cloud.

© Frank Bell, Raj Chirumamilla, Bhaskar B. Joshi, Bjorn Lindstrom, Ruchi Soni, Sameer Videkar 2022
F. Bell et al., *Snowflake Essentials*, https://doi.org/10.1007/978-1-4842-7316-6_9

SQL Basics

SQL stands for Structured Query language and was first introduced in 1974. SQL was designed to manage the data present in RDBMSs. SQL can be categorized further as

- Data Definition Language (DDL): DDL statements are used to define, change, or delete the structure of objects. The following are DDL examples for a database table:

```
CREATE TABLE <TableName>
ALTER TABLE <TableName>
DROP TABLE <TableName>
```

- Data Modification Language (DML): DML statements are used to query or manipulate the data inside the database object. The following are examples of DML statements:

```
SELECT ... FROM <TableName(s)> WHERE ...
INSERT INTO <TableName> VALUES ...
UPDATE <TableName> SET ... WHERE ...
DELETE FROM <TableName> WHERE ...
```

- Data Control Language (DCL): DCL statements are used to control the access to the data in the RDBMS or the database object. Snowflake examples covered in Chapter 6 are DCL. In ANSI SQL, you can define following statements as DCL:

```
GRANT SELECT,INSERT,UPDATE,DELETE on <TableName>
REVOKE SELECT,INSERT,UPDATE,DELETE on <TableName>
```

Now that we have seen broad categories of ANSI SQL statements, let's look at how you can get started and write the first select statement in Snowflake. In Snowflake, we can format a simple select SQL as in the following:

```
SELECT <Column_Names>
FROM <TableName>
WHERE <Filter_Condition>
```

The SELECT clause defines which columns will be part of your query result output. The FROM clause specifies the table or view name(s) from which the SQL will select the data records, and the WHERE clause defines the condition that data records must satisfy.

The WHERE clause acts like a filter. Figure 9-1 shows a code example of simple SELECT, FROM, and WHERE clauses in Snowflake.

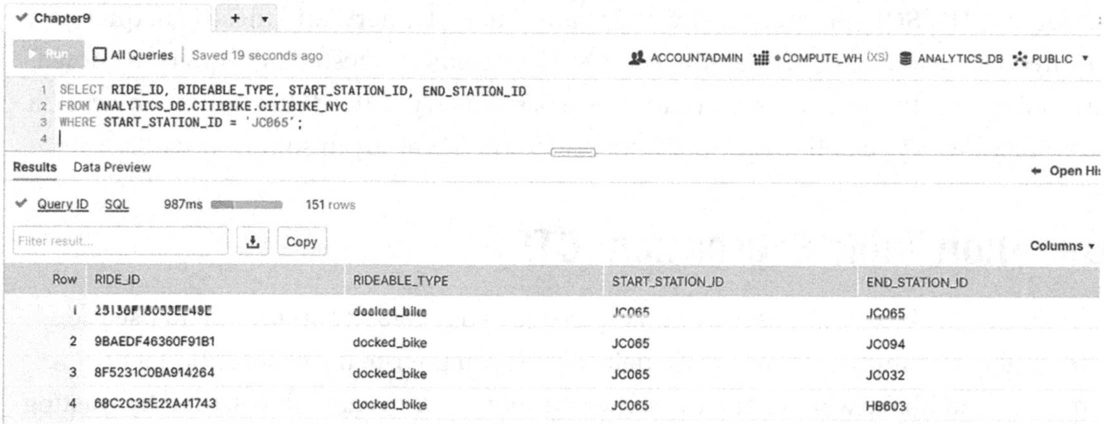

Figure 9-1. *SELECT Example Statement in Snowflake*

Now let's have a quick overview of different Snowflake SQL clauses available to work on more advanced data use cases.

Joins

The data we need is sometimes not present in one single table. In such cases, we need to combine, connect, or join two tables based on a common column(s) and select the columns we need. There are five types of joins, and each of these JOINS works differently and returns different results.

In the following, you can see the usage of INNER JOIN, LEFT OUTER JOIN, RIGHT OUTER JOIN, FULL OUTER JOIN, and CROSS JOIN:

```
SELECT
<Table1_Column_Names and Table2_Column_Names>
FROM
Table_Name1 [INNER|LEFT OUTER|RIGHT OUTER|FULL OUTER|CROSS]
JOIN Table_Name2 on <Join_Condition(s)>
WHERE
<Filter_Condition(s)>
```

Subquery

You can use the output of one SQL query into another SQL using a concept of a subquery. The SQL query can be nested in another SQL query, and these SQL queries are connected via a relational operator, IN or NOT IN clause. A nested SQL query is known as a subquery. In most cases, you can rewrite a subquery with a JOIN clause; however, a subquery helps you define the necessary execution hierarchy in some use cases.

Common Table Expression (CTE)

Continuing with our subquery examples, let's say you have a requirement to use the *same* subquery multiple times in a single SQL. If you rewrite the exact subquery n times, the database will execute the same subquery n number of times, thereby wasting resources and taking more time. It is also cumbersome to make a common change in all the subqueries. Common Table Expression (CTE) helps us to mitigate these challenges. Using a WITH clause, you can define a Common Table Expression (CTE) and then refer to the CTE's output as a table in your main SQL query.

Tip   You can have multiple WITH clauses before SELECT. However, the WITH clause cannot contain forward references. Furthermore, you can use the WITH clause in most of the DMLs but not (yet) in DDLs.

CTE use is also highly recommended to handle table self-joins and recursive joins (mostly used in hierarchy data cases). CTE reduces the complexity and makes the code more maintainable.

The WITH clause comes before the SELECT clause. In the following code example, you can see the advantage CTE offers:

```
WITH CTE_Sample as
  (
  SELECT AVG(x) as Needed_Avg_Value
  FROM table4
    [Join Condition(s) with <tables>]
  WHERE
    <Filter_Condition(s)>
  )
```

```
// From Table 1, we need records where C1 is same as
// Needed_Avg_Value
SELECT c1, c2
  FROM table1 WHERE c1 = (SELECT Needed_Avg_Value
                                FROM CTE_Sample)
UNION

// From Table 2, we need records where C1 is higher than the
// Needed_Avg_Value
SELECT c1, c2
  FROM table2 WHERE c1 > (SELECT Needed_Avg_Value
                                FROM CTE_Sample);
```

GROUP BY and HAVING

You can use the GROUP BY clause to apply an aggregate function on a group of records. The GROUP BY clause defines the condition of grouping these records into one single record per group. The HAVING clause is applied on aggregate functions and used to filter grouped records.

Tip Both HAVING and WHERE clauses filter the records but at different stages. The WHERE clause filters records for the GROUP BY clause to be applied, and the HAVING clause is then applied on these already grouped records.

Time Travel

In Snowflake, you can query a historical state or data for database objects like tables. How far can you go back and query depends upon the value in the DATA_RETENTION_ TIME_IN_DAYS object parameter. You can use AT|BEFORE keywords to define the historical timestamp on which your query should run.

ORDER BY

Using the ORDER BY clause, you define how the query result should be ordered.

In addition to the ORDER BY keywords, you need to define the column or columns on which ORDER BY is applied and the order option – ascending or descending.

MATCH_RECOGNIZE

An advanced feature in Snowflake, MATCH RECOGNIZE allows you to define a pattern and select the records from a table that match the defined pattern. For example, you can define patterns to identify spending behavior or website page usage.

Your SQL query in Snowflake can be generalized and formatted as in the following:

```
[ WITH ...]
SELECT
    [ TOP <n>]
    ...
[ FROM ...
    [ AT | BEFORE ...]
    [ CHANGES ...]
    [ CONNECT BY ...]
    [ JOIN ...]
    [ MATCH_RECOGNIZE ...]
    [ PIVOT | UNPIVOT ...]
    [ VALUES ...]
    [ SAMPLE ...]]
[ WHERE ...]
[ GROUP BY ...
    [ HAVING ...]]
[ QUALIFY ...]
[ ORDER BY ...]
[ LIMIT ...]
```

Using SQL in Snowflake Worksheets

As mentioned previously, the Snowflake Worksheets is the main area where you can easily start entering and executing SQL queries in the Snowflake Data Cloud. Remember, you always need to set the context within a Snowflake worksheet before you can execute your queries in Snowflake. Also, besides DDL SQL, you will need to always have a warehouse set for queries to run.

Setting the Context of Your SQL Queries

Snowflake needs to understand the database, schema, and tables we are referring to in our SQL. In the Snowflake web UI, you can set the context by choosing the role, warehouse, database, and schema in the worksheet. However, if the query spans multiple databases or multiple schemas, you need to provide a completely qualified table or view name in your SQL query.

In the Snowflake Classic Console, you can set the role in two places, the worksheet area and the overall role of the interface in the top-right corner of the UI. The first setting at the worksheet level sets the context for your queries and determines what objects you may use from the left-hand-side pane. Each worksheet has its own separate session and has its own role. The role under your username in the top right-hand corner determines the visibility of the top row of icons and also the behavior of the account-level working areas such as Databases, Warehouses, and History. You can set both these roles independent of each other. Figure 9-2 shows an example of the worksheet context set as the ACCOUNTADMIN role, COMPUTE_WH warehouse, ANALYTIC_DB database, and CITIBIKE schema.

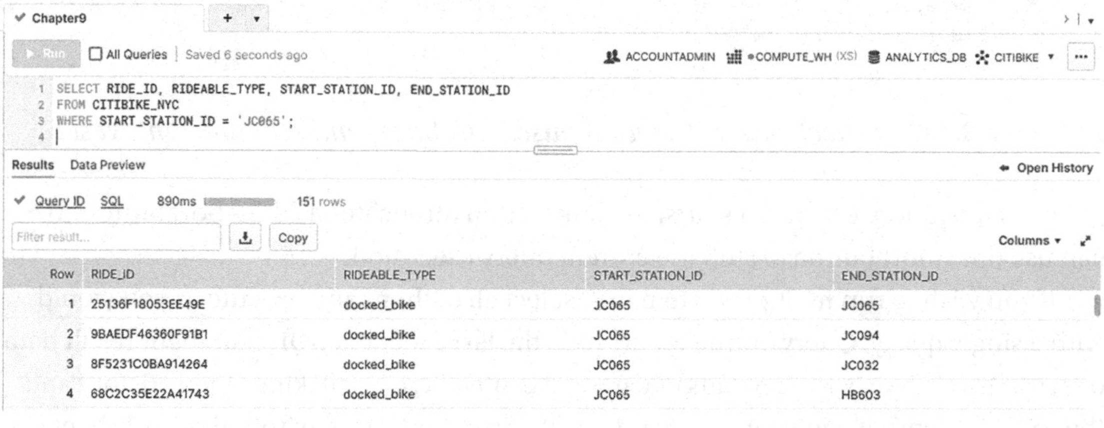

Figure 9-2. *Context Set in Snowflake*

Tip   Before you run any SQL query (DDL or DML), make sure that you have selected the correct role in the worksheet. Creating or querying objects with the wrong role may lead to errors or even compliance issues.

Once you have selected the role for the worksheet, we can check the objects available for the chosen role to query on the left-hand-side pane.

Executing SQL Queries

Once we have checked and set the context, we can now write our SQL queries and execute them in the worksheet's query editor. To execute a single query, we can place the cursor anywhere within the SQL query and click the Run button or use Ctrl+Enter. Figure 9-3 shows an executed query with the worksheet context set in the upper right menu with the role set to ACCOUNTADMIN, the warehouse set to COMPUTE_WH, the database set to ANALYTICS_DB, and the schema set to CITIBIKE.

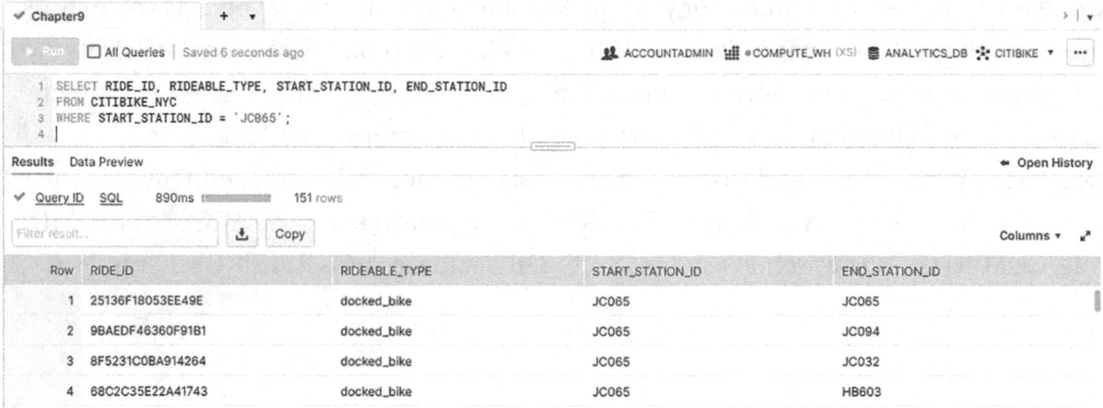

Figure 9-3. *Worksheet with role, warehouse, database, and schema context set*

Once the query execution starts, the Run button duplicates as the Abort button. We can use this Abort button to stop the current query execution.

If you wish to run multiple statements, select all of them and execute as you would with a single query. Query output shows only the last executed SQL statement result data or error. You can check the results of earlier executed SQL by clicking Open History and then SQL statement. Figure 9-4 shows the SQL history interface within the worksheet on the Classic Console.

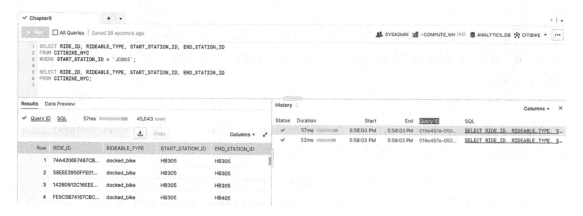

Figure 9-4. *Open History Details in Snowflake*

You can load SQL queries directly from files as well. This is possible by clicking the down arrow at the end of the worksheet tab and then selecting *Load Script*.

Unfortunately, the Snowflake Classic Console does not provide a fail-proof way to save worksheets. The Classic Console worksheets are saved in the user stage and can be accidentally deleted when the user executes a remove command for Internal Stages. Snowflake's new web user interface, Snowsight, has much better features for worksheets. The largest improvement is that Snowsight provides both autosuggest and autocomplete features in the Snowsight worksheet. Snowsight also has versioning built into the worksheets. Figure 9-5 shows an example of Snowsight worksheet history.

How to Use Standard SQL Window Functions

Snowflake also supports the usage of window functions in a SQL query. In its basic form, a window function works on a set of records (window) that are grouped by a condition or an expression.

For example, if we want to calculate sales and rank for each store in each region, we can use an aggregate function in SQL. This window function would aggregate the entire dataset and return only sales per store in each region.

Since we want to calculate sales and then rank them, we can use Snowflake window functions.

Tip   Be extremely careful with NULL value records when working with Snowflake's window functions. Some functions will ignore records with NULL completely.

You can further classify window functions as generic window functions, ranked window functions, and window frame functions:

- Generic Window Functions: SQL functions like average (AVG), sum (SUM), and count (COUNT) are some examples of generic window functions. They operate on a complete query data or the complete window.

- Ranked Window Functions: These functions include RANK(), ROW_NUMBER(), and DENSE_RANK(). Such ranked window functions operate on a subset or on a window frame of a query data. Additionally, ranked window functions will always have an ORDER BY clause.

- Window Frame Functions: In these functions, there is no fixed window frame but a sliding frame. These functions are useful when we need to calculate a running sales total or a moving average. It usually involves querying the current row and rows before and after.

We can find a complete list of Snowflake window functions at `https://docs.snowflake.com/en/sql-reference/functions-analytic.html`.

Nonstandard SQL

In Snowflake, we can define a column as a VARIANT data type, which can hold semi-structured data in JSON format. Querying this semi-structured data is simple and intuitive. We cover how to use and query semi-structured data and the VARIANT data type in depth in Chapter 11.

Advanced Worksheet Sharing and Collaboration with Snowsight

In Chapter 5, we covered Snowsight (Preview App), Snowflake's new web user interface, and its features in detail. In this section, we will demonstrate how easy it is to use Snowsight to query data, import SQL scripts, and use worksheets to collaborate with your team.

In Snowsight, you can open a new worksheet by clicking the "+ Worksheet" button. Besides the "+ Worksheet" button, you have further options to import worksheets from your Classic Console. To open an existing worksheet, navigate to one of the worksheet tabs of *Recent, Shared with me, My Worksheets,* or *Folders*. Figure 9-5 shows these tabs in Snowsight.

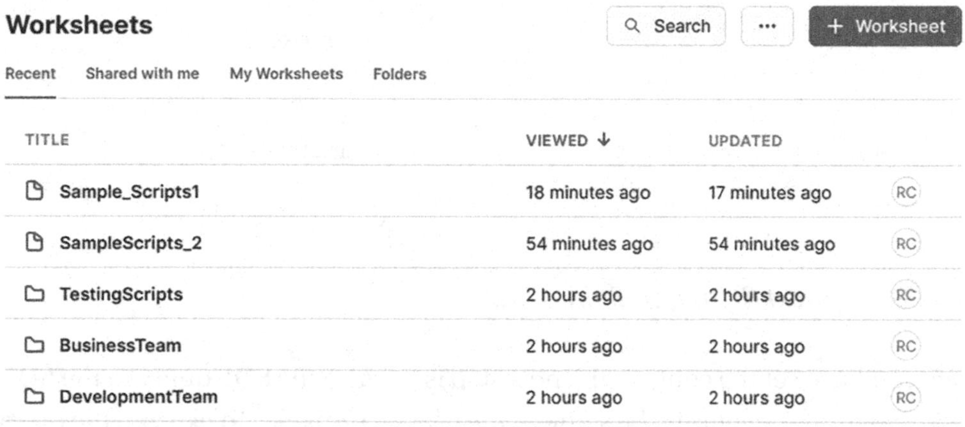

Figure 9-5. *Worksheet Tabs in Snowsight*

Let's quickly review what each tab does:

- **Recent:** List of all worksheets and folders recently used by you.

- **Shared with me:** List of all worksheets shared with you. Snowsight offers a way to share worksheets, folders, and dashboards with other users in a Snowflake account.

- **My Worksheets:** List of all the worksheets created or owned by you.

- **Folders:** List of all folders owned by you or shared with you.

Similar to the Snowflake Classic Console, before you execute any SQL queries, you need to set the query context by selecting the role, warehouse, database, and schema for the worksheet. Remember, each worksheet is in itself a separate session. You can choose a different context for different worksheets. You can set the role and warehouse in the top right-hand corner and select the applicable database and schema in a dropdown above the SQL query editor. Figure 9-6 shows the context set for a SQL execution in Snowsight.

Figure 9-6. *Context Setting in Snowsight*

Once you have set the context, the next step is to write the SQL query in the SQL editor pane. The SQL syntax is the same as the Classic Console. To execute the query, select the SQL or position the cursor anywhere in the query and click the Run button or press Cmd+Return on Mac or Ctrl+Enter on Windows. Figure 9-7 shows the execution of SQL in Snowsight.

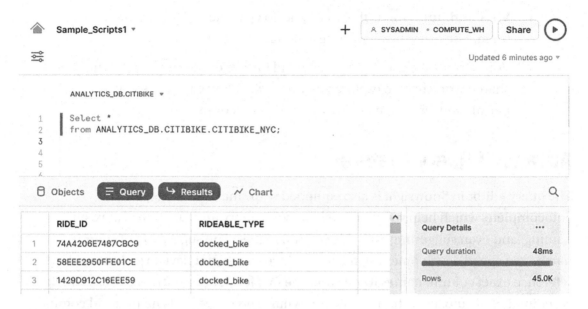

Figure 9-7. *SQL Execution in Snowsight*

Similar to the Classic Console, you can select multiple statements and execute them if you wish to run them together in a sequence. In Snowsight, only the last SQL statement status, that is, the SQL result data or error, is shown in the query output pane. You cannot check previous executed SQL results or SQL history like in the Classic Console. However, you can always use the QUERY_HISTORY view to review the past processed SQL statements. Snowsight as covered in Chapter 5 also allows you to create chart visualizations of the SQL output.

Worksheet and Dashboard Sharing

Snowsight enables you to share your worksheets, folders, and even dashboards with other individuals and teams using your Snowflake account. You can do so by sharing the link to your worksheets, folders, and dashboards with others.

You can set the level of control on the shared worksheet as

- **View Results:** Grants the user rights to view the SQL and the applicable results when the worksheet was shared. The user cannot refresh the results. However, the user can duplicate the worksheet and execute the queries using their role and warehouse.

- **View and Run:** Grants the user rights to view the SQL and execute them. The user must have access similar to the set context.

- **Edit:** Grants the user rights to edit and further enhance a copy of the shared worksheet to explore other use cases. You can only set this permission when you invite an individual account user.

Advanced Query Features

The query editor in Snowsight is also equipped with smart autosuggest and autocomplete, which help you as a developer with query and function syntax, alias naming, and even suggesting available column names. The suggestion of column names can speed development significantly since you do not have to go back and reference exact column names if you know the first few letters. Furthermore, Snowsight's combination of autosuggestion of function syntax and usage hints helps avoid common SQL errors, whereas table or alias column and data type hints help write a better query. Figures 9-8 and 9-9 show the autosuggestions for aliases and functions.

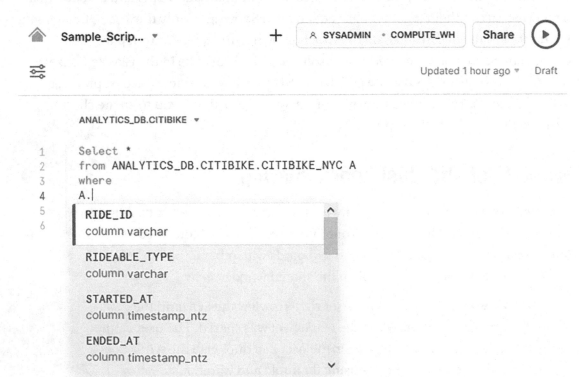

Figure 9-8. *Alias Column and Data Type Hints*

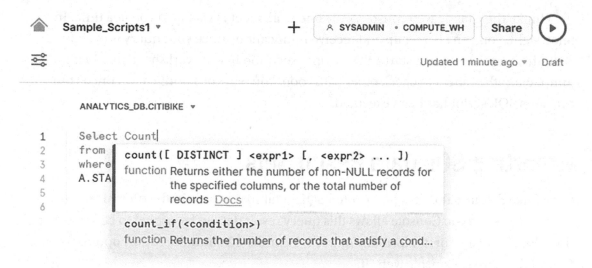

Figure 9-9. *Snowflake Function Autosuggestions*

Snowsight also saves your past versions of the worksheet. We can select a previous worksheet version by clicking Updated <Timestamp> and selecting the timestamp version you need. The SQL editor will update and now show the SQL used or executed at the specified timestamp. You can re-execute this SQL and go back to your latest worksheet version.

We covered Common Table Expression (CTE) earlier in the chapter. Snowsight gives you a more intelligent way to save a frequently used subquery or list of values as defined keywords. You can use the custom filters feature in Snowsight to define these keywords and filters. Your ACCOUNTADMIN should first enable the usage of custom filters for you. Once created, you and other authorized account users can refer to the keyword in SQL queries. Figure 9-10 shows the usage of custom filters in the Snowsight worksheet.

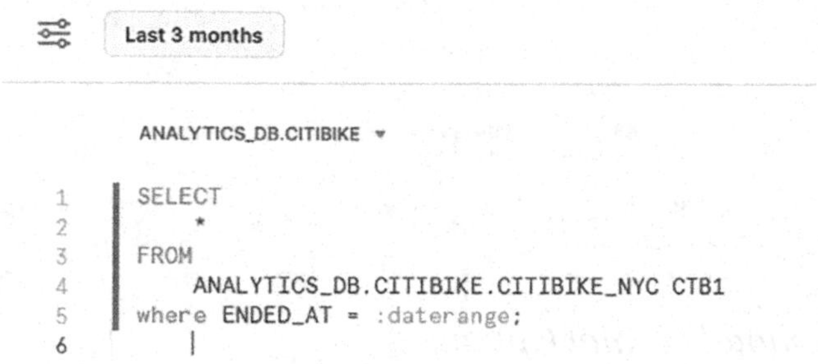

Figure 9-10. *Usage of Custom Filters*

Unlike the Classic web console, only one worksheet is visible to you at a time. In Snowsight, you can also set up and receive notifications once your query is executed. With this feature, you can start a SQL script execution in one worksheet and, in parallel, jump to another sheet to check other SQL code. You will get a notification once your previous SQL script has been executed.

Exporting SQL Data Result Sets

Snowflake displays the data of executed SQL in tabular format in the query result window. The Classic Console allows this query result data to be copied to clipboard or downloaded in .csv or .tsv format. Figures 9-11 and 9-12 show the export options on the Snowflake Classic Console web UI.

Figure 9-11. *Download and Copy Options*

Figure 9-12. *Download Format Options*

By default, Snowflake limits the download size in the Classic Console to 100 MB. Snowflake has implemented this check to safeguard against the risk of the Snowflake user's browser running out of memory, which could cause the browser to crash.

On the other hand, Snowsight currently does not have any such limitation. In Snowsight, there is also no straightforward Copy button, but you can click the leftmost top cell to select the query result and copy-paste it. However, in both consoles, the clipboard size limits the number of records copied.

Query Profile Overview

We have seen how to set the context and execute a query in Snowflake. When you submit a query for execution, Snowflake will compile, optimize, and break the SQL into executable nodes and run it. These details and more are available in the query profile of the query execution.

As a Snowflake user, you might want to analyze your SQL to recognize the execution sequence, heavy operations in SQL, optimizations applied, and even bottlenecks. In the Classic Console, you can first get a high-level glimpse of your query profile right above the query result window. The query ID is a unique ID provided to the query. SQL hyperlink points to the executed query itself, and between the execution time and result count are the query statistics. Figure 9-13 shows the query details available to the user.

Figure 9-13. *Query Execution Details*

The query statistics are color-coded to provide quick details about the execution. You can hover over the query statistics to get further execution details. Figure 9-14 shows typical execution time details for an executed SQL query.

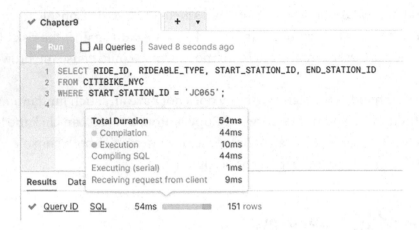

Figure 9-14. *Execution Time Details*

However, if your SQL was executed with an inactive warehouse, you might see additional details as shown in Figure 9-15.

Figure 9-15. *Execution Time Details with Inactive Warehouse*

You can analyze the query plan further by clicking query ID and analyzing the query profile itself. Let's use a bit more complex SQL for our query profile analysis. We will use the following SQL:

```
SELECT START_STATION_ID, END_STATION_ID, count(RIDE_ID) as Ride_Count
FROM CITIBIKE_NYC
WHERE START_STATION_ID = 'JC065'
group by START_STATION_ID, END_STATION_ID
having count(RIDE_ID) > 5
order by count(RIDE_ID) desc;
```

Figure 9-16 shows the SQL and query ID associated with the preceding query, and Figure 9-17 shows the query execution details under the query profile tab.

Figure 9-16. *Executed SQL and the Associated Query ID*

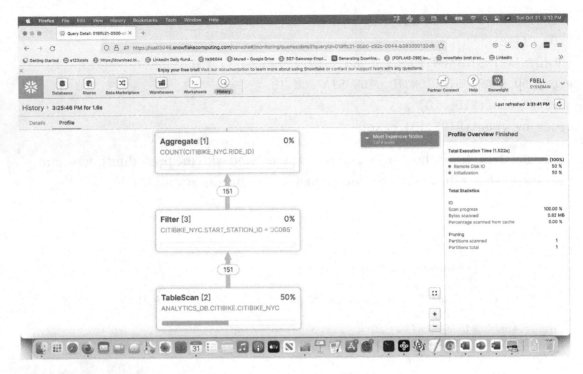

Figure 9-17. *Details Under the Query Profile Tab*

Let's go over some of the details described in Figure 9-17. As you can see, Snowflake further divided our SQL query into executable nodes and executed each of them in a particular sequence. The graphic resulting from the execution sequence is also called the Operator Tree.

Each node's operation result acts as an input for the next node. In our example, Snowflake scanned the table CITIBIKE_NYC first, then applied the WHERE clause filter of START_STATION_ID = 'JC065', aggregated the records for COUNT, used the HAVING BY clause of count(RIDE_ID) > 5, and finally sorted this intermediate result with ORDER BY and displayed the results.

You can also track the record count as it goes through each node. The execution started with the scan of the CITIBIKE_NYC table with 45K records and ended with a sort operation resulting in just six records. These operator counts can clearly show a poorly designed join operation (resulting in more records than expected).

The percentage value in each node indicates the percentage of execution time used by this operator node in the overall execution. On the right-hand side, you see the profile overview information regarding execution time and detailed statistics. The execution time category describes on which operation/phase Snowflake spent time for query execution. This category may list and provide details on the following characteristics:

- **Processing:** The percentage of time spent on actual processing of the data for query execution

- **Local Disk IO:** The percentage of time spent while the execution was blocked by local disk access

- **Remote Disk IO:** The percentage of time spent while the query execution was blocked by remote disk access

- **Network Communication:** The percentage of time spent while waiting for the data transfer over the network

- **Synchronization:** The percentage of time spent while synchronizing execution or processing activities

- **Initialization:** The percentage of time spent for setting up the query for execution or processing

Detailed statistics provide further details about the query, and this category may list and provide details on the following characteristics:

- **IO:** This group provides details about the input-output operations performed.

- **Pruning:** This group provides details about the partition usage to complete the operation. One of the values, "Partitions scanned," denotes the number of micropartitions read and should be as less as possible.

- **Spilling:** This group provides details about disk usage for operations when intermediate results no longer fit in warehouse memory. Always try to design your queries so that there is no spillage at all. If the spillage is not avoidable, then all values in this group should be as less as possible.

Query History

You can view the executed query history at three places in Snowflake:

- **Worksheet Query History:** In Classic Console web UI, the query history for the worksheet is available in the same pane as query results. This query history is only visible for the worksheet owner.

- **History Tab (Near the Worksheets Tab):** In the History tab, you can see the SQL executed at the Snowflake account level. You need to have monitor permissions to see the History tab details. Here you can also apply filters like user, warehouse, and SQL text on the executed SQL metadata.

- **QUERY_HISTORY View:** This is ACCOUNT_USAGE view that can be used to query the metadata about executed SQL in the past 1 year. Here again, you need proper permission to read the data from this view.

You can also assign a query tag to a session or set of queries to easily identify them later. You can easily do so by using the following command:

```
ALTER session set QUERY_TAG = 'Chapter 9';
```

Query tags can easily help you analyze the exact session or a particular data workload related to SQL execution. Figure 9-18 shows the query history details in the History tab by using the query tag filter.

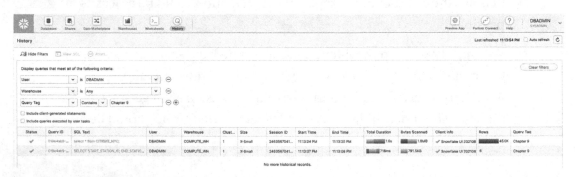

Figure 9-18. *Query Details When Using a Query Tag*

Cloning

Today's data platforms are bigger, faster, and cheaper. Expectations from a data platform are also rising. Modern data platforms are expected to do more than just process and store data. We expect the data platform to give the users the flexibility to explore and try out new implementations without being constrained by time, cost, or database administrators!

Snowflake introduced a feature called Zero Copy Cloning. With the Zero Copy Cloning feature, you can create a copy of a database, schema, or table on demand. *Zero Copy* means that the copy action is only logical. There is no physical data copied or duplicated. The original and cloned objects are independent and can be worked upon separately without impacting each other. Since the object and the data copy are only logical in nature, the cloning operation is fast (not instantaneous) with no additional storage costs or limitations on the number of clones.

From previous chapters, we know the Snowflake Data Cloud stores the data as immutable micropartitions. At any given moment, a set of micropartitions constitutes the current state of the database. Snowflake makes a snapshot of current micropartitions during the cloning operation and presents it as a cloned object.

Snowflake's Zero Copy Cloning gives Snowflake users incredible flexibility and freedom. Users could use cloning features for

- Performing backups

- Testing on cloned data objects that are exactly the same as production

- Enabling object rollback capability for releases

- Creating a free sandbox to enable parallel use cases and exploration

- Creating new DEV and UAT environments on demand

Note   Cloned objects do not incur storage costs as long as the data in cloned objects is not changed, added, or deleted. Newly created and added objects will incur storage costs.

Cloning Syntax

Now that we have discussed how Snowflake handles cloning, let's have a look at the syntax for creating an object clone. Figure 9-19 shows how we can create a clone in Snowflake:

```
CREATE TABLE ANALYTICS_DB.CITIBIKE.CITIBIKE_NYC_CLONE
CLONE ANALYTICS_DB.CITIBIKE.CITIBIKE_NYC;
```

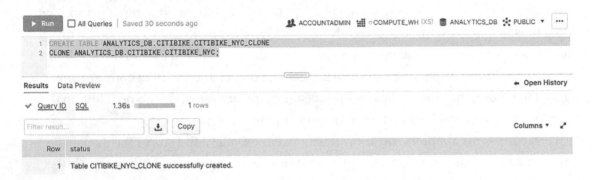

Figure 9-19. *Cloning an Existing Object*

This command creates a cloned table named CITIBIKE_NYC_CLONE, which is the exact copy of the table CITIBIKE_NYC.

The cloning syntax can be generalized as follows:

```
CREATE [ OR REPLACE ] { DATABASE | SCHEMA | TABLE | STREAM | STAGE | FILE
FORMAT | SEQUENCE | TASK} [ IF NOT EXISTS ] <Target_Object_Name>
  CLONE <Source_Object_Name>
```

At the database and schema level, the cloning automatically applies to objects under them. When you clone a database, Snowflake also clones the database's schemas, tables, and other objects in the database. Similarly, when you clone a schema, all contained schema objects are cloned with it. There are, nevertheless, exceptions to this rule. The cloning command will not clone external tables and Internal Named Stages even if they are part of the source database or schema.

Cloning with Time Travel

It is possible to combine the cloning operation with the time travel feature of Snowflake. You can create a clone of a table as per its past state. The time travel feature provides you the ability to access a past state of the table. The account-level parameter DATA_RETENTION_TIME_IN_DAYS defines and influences the time travel feature and how far back it will go. This parameter also has an impact on your storage costs.

Note   While cloning with time travel, only objects and child objects valid at a specified timestamp are cloned.

You can combine cloning and time travel for a table as in the following:

```
CREATE [ OR REPLACE ] { TABLE } [ IF NOT EXISTS ] <Target_Clone_
Object_Name>
  CLONE <Source_Object_Name>
[BEFORE|AT (
        [TIMESTAMP => <Required_TimeStamp> |
        STATEMENT => <Query_ID>]
]
```

Snowflake will throw an exception during cloning

- If you provide an invalid time travel timestamp for the source object

- If the object did not exist at the specified timestamp

Cloning Considerations

While executing cloning operations, you need to be mindful of certain considerations and impact of parallel SQL commands:

- **Impact of DDL:** Cloning an object like a database and schema takes time. During this period, if you drop or rename a table, Snowflake might throw an error as it can no longer find the table to clone.

- **Impact of DML:** The cloning operation may be disrupted and fail, if a DML command like delete or truncate table is executed, thereby deleting the data to be cloned.

221

- **Behavior of Cloned Objects like Pipes, Streams, and Tasks**: Pipes that work with an Internal (Snowflake) Stage are not cloned, whereas the ones referencing External Stages are cloned. With streams, any unconsumed stream records are not available in the cloned object. Tasks are by default cloned but are in suspended state. You need to resume them by using the ALTER command.

- **Privileges on Cloned Objects:** Only if a database or schema is cloned the child objects retain their granted privileges. Else, a cloned object like a table does not retain any of the source object's original granted privileges.

Summary

Snowflake supports ANSI SQL and other native-built functions to help you navigate any data query demands. Snowflake optimizes and breaks the given SQL into executable nodes. Details of the execution pattern and nodes are available in the query profile tab. We can analyze this execution profile and recognize any bottlenecks in the processing of the query.

Using available query history options, you can also quickly analyze the previous SQL executions or Snowflake consumption or even troubleshoot any data issues. However, your user needs to have proper privileges to access the query history metadata.

Snowflake further provides "Zero Copy" clones, which are logical copies of the source object and do not incur storage costs. You can also combine Snowflake's time travel and cloning features.

Snowsight is Snowflake's new web UI console and encourages collaboration via worksheet and dashboard sharing features. Custom filters and SQL autosuggest features are also beneficial for citizen business users.

How Snowflake Compute Works

In this chapter, we will cover how compute works within the Snowflake Data Cloud with Snowflake's compute warehouses. As mentioned previously, the separation of compute from storage was a key architectural foundation of the Snowflake Data Cloud. When Snowflake was founded back in 2012, this was an entirely revolutionary concept. Prior to Snowflake, no database technology had ever separated storage from compute, so to this day this still gives Snowflake pay-as-you-use data processing with cost advantages. The Snowflake Data Cloud allows you to enable and control the underlying cloud provider compute resources through their warehouse construct.

One very key point of clarification around Snowflake warehouses is they are only related to compute resources. This can be very confusing if you are coming from an on-premise database or working with data warehouses for many years. You must move your paradigm from warehouses being these entities of storage and compute to ONLY being virtual warehouses or virtual compute. We will cover the Snowflake warehouses in much more depth in this chapter, but it can help to think of Snowflake warehouses as flexible configurations that activate compute resources on the cloud provider that the Snowflake account is operating on.

Snowflake Compute Warehouses
How Snowflake Warehouses Really Work

Snowflake warehouses make up the compute layer of Snowflake. Remember, though, that Snowflake warehouses again are just configurations of the amount of compute to deploy when a warehouse is resumed or running. Technically, you could create a Snowflake account and never ever execute a job and never engage the compute

© Frank Bell, Raj Chirumamilla, Bhaskar B. Joshi, Bjorn Lindstrom, Ruchi Soni, Sameer Videkar 2022
F. Bell et al., *Snowflake Essentials*, https://doi.org/10.1007/978-1-4842-7316-6_10

layer. This Snowflake compute layer with its virtual compute warehouses provides the compute power to process data within the Snowflake Data Cloud. Snowflake warehouses are really the main processing and working layer of raw compute resources. Any new query (SELECT) or Data Manipulation Language (DML) including UPDATE, INSERT, DELETE, MERGE, etc. running on Snowflake will require a compute warehouse to be running for the data to be processed. The only exceptions are when the exact query result has been cached or the SELECT statement is querying metadata.

Under the hood of a Snowflake Data Cloud warehouse is a set of virtual servers in a cluster. (Remember, clusters though are only available for the Snowflake Enterprise edition and above). Each virtual server comes with its own CPU, memory, and disk storage in SSD drives. Figure 10-1 shows the Snowflake compute layer in the overall three layers of the Snowflake Data Cloud.

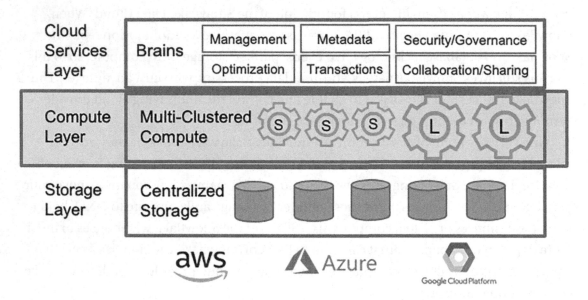

Figure 10-1. *Snowflake Compute Layer (also known as Multi-cluster Compute Layer)*

This compute layer consumes Snowflake credits when a warehouse is resumed to process jobs. Warehouses that are in *suspended* mode can go into *running* mode with very little latency as long as the *Auto Resume* parameter is set to true. By setting *Auto Resume* to be automated, then you create a completely on-demand compute resource solution that turns on when you send jobs to the Snowflake Data Cloud. This compute resource processes the job(s), and when it's done it uses the settings related to warehouse *Auto Suspend*. This *Auto Resume* feature really is one of the key parts of the

Snowflake Data Cloud flexibility that enables this amazing capability to go from having no compute resources active to enabling up to 512 virtual machines to process your data operations. You can go from not paying anything to just paying for warehouse compute resources you actually need. If you came from an on-premise database or other cloud databases, this is an incredibly amazing capability.

The coordination related to turning on the warehouse is performed and controlled by the Snowflake Data Cloud services layer, which will have the warehouse running even after the execution job that required its services is completed. This also means that the warehouse will continue to use the Snowflake credits after the execution job has been completed depending on what is set with the *Auto Suspend* setting. Snowflake does have guardrails built in; after Snowflake detects a preconfigured time of inactivity in the warehouse related to the *Auto Suspend* setting, Snowflake will spin down the virtual machine(s), thus saving unnecessary consumption of Snowflake credits. Snowflake credits are how Snowflake bills for using both compute resources and cloud services. The compute credit meter is turned on to track the warehouse usage and tracked per second with one-minute minimum.

To give an example of how a Snowflake warehouse works when a query is issued to Snowflake, let's illustrate a sequence of events:

- A query is issued to Snowflake with a warehouse that is currently in the *suspended* state.

- Snowflake starts up the warehouse (goes into *running* state), and the query is executed by the running warehouse.

- The time between the query issue time and the query start time is named queuing.

- For this example, let's assume the query completes in 10 seconds.

- For this example, let's also assume that this warehouse has the *Auto Suspend* parameter set to 10 minutes. Due to this parameter, the warehouse will stay in the running state for 10 more minutes before going to the *suspended* state, that is, provided that no other query is issued to this warehouse during that 10 minutes.

The total time charged will be 10 minutes and 10 seconds if no other queries are executed.

> **Tip**   Remember, Snowflake warehouses are ONLY configurations or pointers
> related to compute resources on the underlying cloud provider. Snowflake
> warehouses have nothing to do with conceptual data warehouses that include
> storage.

Snowflake Warehouse in the Snowflake UI

Figure 10-2 shows an example of these configurations on the Snowflake Classic Console
Create Warehouse creation screen. A warehouse can be created and managed in the
Snowflake UI or through code.

Snowflake UI for the Creation and Management of a Warehouse

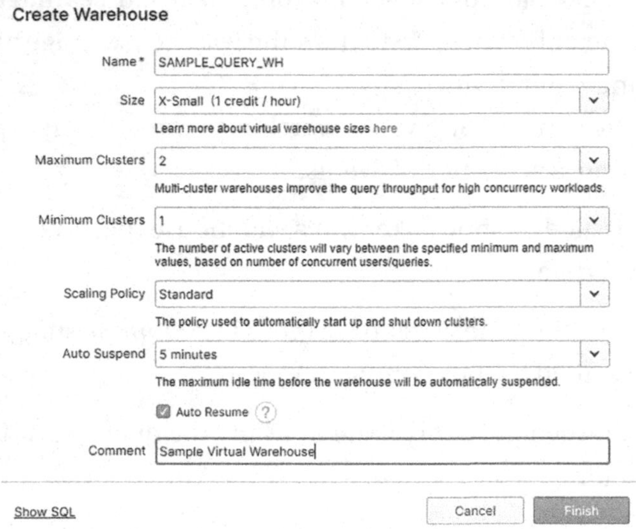

Figure 10-2. *Create Warehouse Example – Classic Console*

Sample Code for the Creation and Management of a Warehouse

```
CREATE WAREHOUSE "SAMPLE_QUERY_WH"
SET WAREHOUSE_SIZE = 'XSMALL'
AUTO_SUSPEND = 300
AUTO_RESUME = TRUE
MIN_CLUSTER_COUNT = 1
```

```
MAX_CLUSTER_COUNT = 4
SCALING_POLICY = 'STANDARD'
COMMENT = 'Sample Virtual warehouse';
```

Snowflake Warehouse Sizes

Snowflake warehouses are designed to *scale up* and *scale down*. They start from an extrasmall (XS) cluster with a single virtual server all the way to a 6 extralarge (6XL) cluster with 512 virtual servers. With each increment in T-Shirt sizing of the warehouse, the cluster doubles in size, providing double the raw compute power. This also means the credit consumption rate between any given two increments doubles as well. The following chart shows the correlation of Snowflake T-Shirt sizes with the number of servers in a warehouse and the number of credits the warehouse will consume for each hour the warehouse actively runs.

Warehouse Size	Servers	Credits / Hour
X-Small	1	1
Small	2	2
Medium	4	4
Large	8	8
X-Large	16	16
2X-Large	32	32
3X-Large	64	64
4X-Large	128	128
5X-Large	256	256
6X-Large	512	512

Snowflake Multi-clustering

In addition to *scale-up* and *scale-down*, Snowflake warehouses are also designed to *scale out* with their multi-cluster warehouse architecture. A multi-cluster warehouse can spawn additional compute clusters of the same size as the original cluster to manage workloads with additional concurrency needs. The scale-out is controlled by *Minimum Clusters* and *Maximum Clusters* settings as shown in Figure 10-2. The multi-cluster warehouse feature is only available in the Snowflake Enterprise edition and higher. The entry-level Standard edition of Snowflake comes with a standard warehouse with no *scale-out* features.

Multi-clustering Settings

You will notice in Figure 10-2 the additional settings for Snowflake's multi-clustering feature of Minimum Clusters, Maximum Clusters, and Scaling Policy. (Again, these will not be available if you are not using the Enterprise edition or above. Let's go through each of these settings and explain how they work:

- Minimum Clusters: This is the minimum number of clusters that a warehouse starts up with. If you have massive concurrency and don't want any additional queuing to happen when the warehouse cluster is first launched, you can set this to a value of two to ten.

- Maximum Clusters: This is the maximum number of clusters that a warehouse will scale out to.

- Scaling Policy: Currently Snowflake has two options for controlling the scale-out of a multi-cluster warehouse. You can choose either Economy or Standard. Standard is selected by default. Assuming you have additional clusters to scale out to, then the Standard setting turns on the next cluster as soon as it detects queuing. The Economy setting only starts up an additional cluster if it projects that the next cluster will have a minimum of an additional six minutes of active work to perform. You can decide on one of the settings based on if you want almost no queuing or you prefer to spend less on additional compute costs related to new clusters being turned on without six minutes of workload to handle.

Snowflake Compute Strategies

Now that we have looked at Snowflake warehouse sizes and types (standard and multi-cluster), let's see how both can be leveraged to design the correct warehouse compute needs for you. The key to designing Snowflake compute warehouses is to start small, both in *scale-up* and *scale-out* principles, and expand as required to handle more complex queries and data volumes.

Scaling Out

The extrasmall (XS) Snowflake warehouse is capable of handling eight threads by default. When more jobs are sent to Snowflake exceeding the maximum limit, Snowflake will start queuing the jobs as it waits for the running jobs to complete. When queuing like this is detected, it means there are not enough resources to handle the incoming queries. The strategy to handle such concurrency needs in Snowflake is to enable the multi-cluster warehouse setting for scaling out. The multi-cluster can be configured to automatically add up to ten nodes or downsize back to a minimum cluster setting as the workload decreases. This balancing act allows the capability to provide maximum concurrency while intelligently utilizing the compute credits. *Scale-out* is a great feature to have to meet SLA with any peak compute usage that occurs within your environment. Figure 10-3 shows a visualization of how you can scale out or in depending on your multi-cluster settings.

SCALING OUT

Adding more clusters to your
current Virtual Warehouse *without*
changing the size of the Warehouse
to handle concurrency issues.

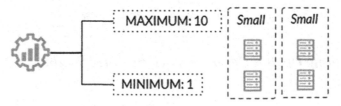

Figure 10-3. *Snowflake Compute – Scaling Out*

Scaling Up: Spilling

Scaling out is meant to solve resource concurrency limitations. Scaling up is for when you need to handle more complex queries or larger datasets. When the complexity of the queries being run increases and more and more data needs to be processed, the queries may start to take longer to run as the warehouse they are running on may not be adequate anymore. The CPU, RAM, and SSD that are available to the smaller warehouse may be overloaded for the query complexity or data size. A telltale sign that a cluster is not adequate in size is when you observe bytes spilling to local disk or the bytes spilling over the network. In order to prevent spilling, you need to *scale up* from a smaller T-shirt-sized warehouse to a larger warehouse, thus adding more CPU, RAM, and SSD as illustrated in Figure 10-4. A warehouse can be altered to the next T-Shirt size to handle more complex queries or larger datasets. This can be done either by the UI or through code by running the ALTER WAREHOUSE command. When a warehouse is altered, if it is in the suspended state, it will start at the new size when the warehouse is resumed next. If the altered warehouse is already in the running state, any queries that are currently running will be allowed to finish at the current size, and any new queries that execute after the resizing of the warehouse will run using the newly sized warehouse.

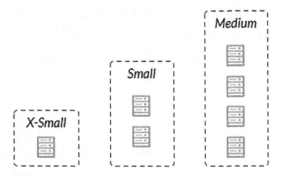

SCALING UP

Resizing a warehouse to handle
complex/process-intensive queries

Figure 10-4. *Scaling Up and Down – Warehouse Size Scaling*

Scaling Up vs. Scaling Out

To summarize, *scale-up* adds raw horsepower to the cluster by increasing its CPU, RAM, and SSD and is to be used to handle more complex queries running on larger datasets. On the other hand, *scale-out* adds additional clusters to the warehouse and is used to handle concurrency needs. It is important to understand both *scale-up* and *scale-out*. Both will double the credits used but will behave differently and are designed for different types of workloads. In Figure 10-5 we can observe that eight credits are consumed per hour, either through one L-sized warehouse or two M (medium)-sized warehouse clusters or four S (small)-sized warehouse clusters or eight XS (extrasmall) warehouse clusters.

CREDITS PER HOUR	Scaling Out For Concurrency - # of clusters									
	1	**2**	**3**	**4**	**5**	**6**	**7**	**8**	**9**	**10**
6XL	512	1024	1536	2048	2560	3072	3584	4096	4608	5120
5XL	256	512	768	1024	1280	1536	1792	2048	2304	2560
4XL	128	256	384	512	640	768	896	1024	1152	1280
3XL	64	128	192	256	320	384	448	512	576	640
2XL	32	64	96	128	160	192	224	256	288	320
XL	16	32	48	64	80	96	112	128	144	160
L	(8)	16	24	32	40	48	56	64	72	80
M	4	(8)	12	16	20	24	28	32	36	40
S	2	4	6	(8)	10	12	14	16	18	20
XS	1	2	3	4	5	6	7	(8)	9	10

Scaling Up and Down — Virtual Warehouse Sizes

Figure 10-5. *Scaling Up vs. Scaling Out*

Monitoring Snowflake Compute (Warehouses)

The simplicity and ease you now have as a data user with Snowflake compute is unprecedented. You can take an execution job with one query running on an XS and change the compute scale on it to an L in milliseconds. You could even take a Snowflake warehouse from a size of XS to a 6XL with a whopping 512 virtual servers in most likely minutes. I have not tested from an XS to a 6XL yet, but when I did it from an XS to 4XL previously, it did take minutes because it depends on Snowflake's backend compute preparation as well as its compute availability algorithm and how many available virtual machines are there. If it's less than the 511 needed to go from XS to 6XL, then it is dependent on how fast those machines could be brought online by Snowflake for your region and your compute usage. Also, remember most likely there are hundreds of other Snowflake customer accounts running in your region. It is entirely possible that another one of the hundreds of customers increases or turns on their compute in the same timeframe, requiring Snowflake to turn on even more virtual servers to meet the demands of its customers.

With this unprecedented data processing compute power also comes great responsibility. If you do not appropriately use automated tools such as Snoptimizer or your own custom-coded or configured solutions, then you are submitting yourself and your organization to significant compute cost risks. By default, Snowflake does not set up guardrail settings on your warehouses. It is up to you the Snowflake administrator or user with Create Warehouse access to do this. I recommend creating a specific resource monitor that has daily credit limits for each warehouse you have. In the following, we will cover what resource monitors are and how they currently work. We will also go over all the options you have to monitor and control Snowflake compute consumption and cost. This is crucial for your Snowflake usage and journey to be a success.

Resource Monitors

Now that we have gone through the extreme cost risks with Snowflake compute with their Snowflake warehouses and the imperative to use automated tools like Snoptimizer or set up your own resource monitors or other monitoring tools, let's go over what resource monitors are as well as how you can set them up.

How to Set Up a Single Resource Monitor: UI and Code

Resource monitors in Snowflake can be set up using the Snowflake web interface or through code. Figure 10-6 shows the initial Create Resource Monitor screen.

Figure 10-6. *Create Resource Monitor – Classic Console*

Set up from code:

```
CREATE RESOURCE MONITOR "TEST_RM" WITH CREDIT_QUOTA = 150
 TRIGGERS
 ON 95 PERCENT DO SUSPEND
 ON 100 PERCENT DO SUSPEND_IMMEDIATE
 ON 80 PERCENT DO NOTIFY;
ALTER WAREHOUSE "TEST" SET RESOURCE_MONITOR = "TEST_RM";
```

Options Available to Monitor and Control Snowflake Compute Costs

There are primarily four options currently available for you to monitor and control your Snowflake compute consumption costs. The four options are as follows:

1. Manually create resource monitors with Suspend enabled.

2. Automate resource monitor creation with Suspend enabled.

3. Custom code – internal or external to Snowflake.

4. Automated services such as Snoptimizer.

Also, while using reporting tools to monitor Snowflake compute costs through account usage views is another option as well, I am ONLY including the preceding options where the warehouses can be automatically suspended if an issue happens and max credit limits are reached. Reporting is an excellent way to monitor, but I strongly feel that this is a suboptimal solution for controlling costs. Some rogue user could create a 6XL, and if you do not catch it within 1 hour with just one cluster, it will cost you 512 credits or $1536, assuming $3 as a standard credit cost. Also, it is incredibly easy to change an existing warehouse compute from XS to L or 4XL by mistake via the GUI if you are going fast. This is why you MUST install credit limits with resource monitors or with scripts that will 100% of the time shut off warehouses based on your threshold for compute cost risk.

Manually Create Resource Monitors with Suspend Enabled

If you are just starting out with Snowflake and only have a few warehouses, then it's pretty simple to manually create a Snowflake resource monitor for those warehouses through the Snowflake web interfaces (Classic Console or Snowsight) or through code. In order to actually control costs, you MUST make sure you enable either Suspend or Suspend After Query. Otherwise, notifications can be missed, and cost risk exposure is much too high.

Automate Resource Monitor Creation with Suspend Enabled

If you have a reasonable number of warehouses or you just want to have more automation around covering yourself and your company against compute cost risks, then you can code a simple set of tasks that will look for any warehouses that do not have a resource monitor and automatically create it and also attach it to the specific warehouse. It is too easy to create a resource monitor and NOT attach it to a warehouse. This basically creates a monitor that is not activated.

Custom Code – Internal or External to Snowflake

Another option that customers with engineering resources take is you can custom code both monitoring and cost control systems either internal or external to Snowflake. You can build monitoring and warehouse cost control through automating suspending of the warehouse(s) once limits are reached that prevents unlimited compute costs from occurring.

Snowflake has internal mechanisms such as tasks, functions, stored procedures, and other objects where you can custom code a combination of these to review warehouse and cost usage within the metadata stats views regularly and take actions when thresholds are reached.

Other Snowflake developers have used many of the external tools such as Python, Java UDFs, lambdas, or their equivalents to similarly monitor credit usage and warehouse consumption as well as take action if thresholds are reached. Just realize with these custom coded monitoring and threshold action solutions, you also have the fun and cost of maintaining this code. Snowflake continues to also optimize its feature offerings, and new solutions with new costs or variations of the existing account usage reporting happen all the time.

Automated Cost Optimization Services such as Snoptimizer

If you are new or an advanced customer but have been burned by many of the cost anti-patterns that can happen and you just want fast automated cost optimization and cost risk minimization, then I recommend the Snoptimizer service. It is the easiest way to automatically optimize your Snowflake account for both cost optimization and cost risk minimization. Of the hundreds of accounts that I have tuned up with optimization and health checks, I always have seen additional cost risk minimization or cost optimization settings that can be improved. Snoptimizer automates all of these within a few hours. It is the only service I've seen of this kind that has helped my clients tremendously. It is incredibly easy to set up as well. You just go here and start your free trial:

```
https://snowflakesolutions.net/what-is-snoptimizer/
```

You can have all the cost optimization and cost risk reduction benefits quickly with this service.

Cost Optimization Best Practice Queries

Before using the following queries, you MUST have access to either the live views or a copy of the Snowflake account usage metadata tables that are on every single Snowflake account.

> **Tip**   The easiest way to get access to these incredibly useful account usage views
> if you do not have the ACCOUNTADMIN is to have your ACCOUNTADMIN execute the
> following statement for your specific role:
>
> ```
> GRANT IMPORTED PRIVILEGES ON DATABASE SNOWFLAKE TO
> ROLE <role>;
> ```

Analyze Credits Used by Warehouse Overall

```
SELECT WAREHOUSE_NAME, SUM(CREDITS_USED_COMPUTE)
FROM ACCOUNT_USAGE.WAREHOUSE_METERING_HISTORY
GROUP BY 1
ORDER BY 2 DESC;
```

Analyze Credits Used by Warehouse – Last 7 Days

```
SELECT WAREHOUSE_NAME, SUM(CREDITS_USED_COMPUTE)
FROM ACCOUNT_USAGE.WAREHOUSE_METERING_HISTORY
WHERE START_TIME::DATE > DATEADD('DAYS', -7, CURRENT_DATE)
ORDER BY 2 DESC;
```

Analyze Credits Used by Warehouse

```
SELECT START_TIME, WAREHOUSE_NAME, CREDITS_USED_COMPUTE
FROM SNOWFLAKE.ACCOUNT_USAGE.WAREHOUSE_METERING_HISTORY
WHERE START_TIME >= DATEADD(DAY, -7, CURRENT_TIMESTAMP())
AND WAREHOUSE_ID > 0
ORDER BY 1 DESC 2;
```

Summary

Snowflake compute is used by creating and resuming Snowflake warehouses. Snowflake warehouses come in T-Shirt sizes from XS (1 node) to 6XL (512 nodes). We covered Snowflake warehouses in depth in this chapter to help you understand how to efficiently use Snowflake compute for your data processing needs. The two standard approaches for scaling Snowflake compute are either to scale up or to scale out. We also covered resource monitors, which are one of the most important features for effectively controlling and optimizing Snowflake compute costs. We hope this provides you with an in-depth understanding of both Snowflake warehouses and resource monitors and how to effectively manage the scale and cost of the Snowflake Data Cloud.

CHAPTER 11

Semi-structured Data in Snowflake

In this chapter, we will cover how the Snowflake Data Cloud handles semi-structured data such as JSON and XML. As data capture and sources have grown significantly in the last few years especially, semi-structured data has also grown. JSON semi-structured data especially grew significantly with the popularity of NoSQL databases. It has become essential for organizations to be able to process semi-structured data and combine it with structured data for analysis. However, analyzing semi-structured data using traditional methods has been more difficult due to added levels of complexity. Many businesses have struggled to combine structured with semi-structured data. Most analytical cloud databases also treated semi-structured data as a second-class data, making it hard to easily use both semi-structured data and structured data.

Snowflake from early stages was designed to handle not only structured delimited file loading and data processing but also files or streams of semi-structured data. Snowflake's unique architecture, which we discussed in Chapter 3, allows semi-structured data to be saved in the database in its natural format. Snowflake also has made it extremely easy to virtually convert semi-structured data into structured views. This is another one of Snowflake's excellent features, which separates it from its competition. Prior to Snowflake, no database technology had ever separated storage from compute. The Snowflake Data Cloud multi-cluster, shared data architecture allows the separation of compute resource scaling from storage resources enabling seamless, non-disruptive scaling. Snowflake's architecture enables consolidation of both structured and semi-structured data into one platform, which enables analytics against a consolidated dataset.

One of Snowflake's early patents is around the concept of a VARIANT data type, which we will discuss in greater depth later in this chapter. This VARIANT data type is at the core of how Snowflake handles semi-structured data. Let's start with understanding how the VARIANT data type works in the Snowflake Data Cloud with some specific examples.

© Frank Bell, Raj Chirumamilla, Bhaskar B. Joshi, Bjorn Lindstrom, Ruchi Soni, Sameer Videkar 2022
F. Bell et al., *Snowflake Essentials*, https://doi.org/10.1007/978-1-4842-7316-6_11

Semi-structured Data Type in Snowflake

Snowflake provides capabilities that make it very easy to handle semi-structured data. This includes flexible schema data types for loading semi-structured data without need for any transformation. Behind the scenes, Snowflake automatically converts semi-structured data to be optimized for internal storage and efficient SQL querying.

Snowflake supports data types that enable businesses to query semi-structured datasets (e.g., JSON and XML) in a fully relational manner. Presently, it provides three unstructured data types of VARIANT, OBJECT, and ARRAY. The VARIANT data type is the most used of these three data types. This VARIANT data type is a universal type, which can store the value of any other type, including OBJECT and ARRAY, up to a maximum size of 16 MB compressed. The VARIANT data type acts as a regular column within a relational table within Snowflake. The ARRAY data type is a list-like indexed data type that consists of variant values. The OBJECT data type consists of key-value pairs, where the key is a Not Null string and the value is a VARIANT type data.

Regardless of the data type specified, data is stored in compressed columnar binary representation for better performance and efficiency. Due to Snowflake's unique architecture and separation of compute from storage, this is completely transparent to users, and they can run queries on these tables like standard relational tables.

How the Variant Data Type Works

Now that we have an overview of semi-structured data types in Snowflake, let's take a deep dive into the VARIANT data type. As the name suggests, this data type can store various types of data. In other words, a value of any data type can be implicitly cast to a VARIANT value, subject to size restrictions. Non-native values like dates and timestamps are stored as strings in a VARIANT column.

Now let's understand the basic operations we can perform on the VARIANT data type. Please log into your Snowflake worksheet using your existing account. Alternatively, Snowflake also provides a free trial account that you can use for 30 days to get started on Snowflake. Let's get started. Figure 11-1 shows the command to create a table named JSON_WEATHER_DATA with only one column named v and the data type as VARIANT. We will be using this table in all our future examples and calculations.

```
Create table json_weather_data (v variant);
```

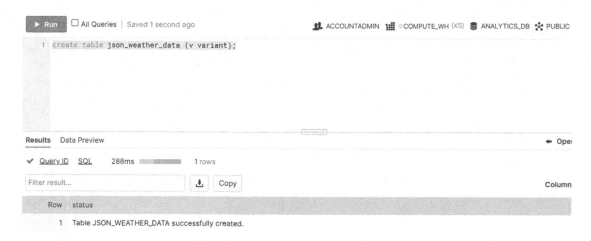

Figure 11-1. *Creating a Table with a Variant Data Type*

Different functions are available to support type casting, handling NULLs, etc. One very useful function for VARIANT data is FLATTEN, which explodes nested values into separate columns to be used to filter query results in a WHERE clause. We will cover this function in detail in subsequent sections.

How to Load JSON Data

JSON (JavaScript Object Notation) is an open-standard data format for semi-structured data stored as nested objects and arrays generated from various sources and devices like web browsers, servers, mobile phones, etc. JSON is text-based, derived from JavaScript, simple to design, and flexible.

Due to the increased popularity of JSON as a very common format for semi-structured data, there was always a need for a data store that supported JSON files. Previously, to query JSON files, users had to load them into JSON-enabled databases, parse data, and then move data to relational databases. The entire process was really time consuming, and this inability to deliver decision support within the shortest time impacted business. Snowflake came as a revolutionary cloud platform to offer native support to load and query JSON files.

We created the table JSON_WEATHER_DATA in our worksheet earlier. Now let's create a stage by running the query in Figure 11-2.

```
create stage nyc_weather url = 's3://snowflake-workshop-lab/weather-nyc';
```

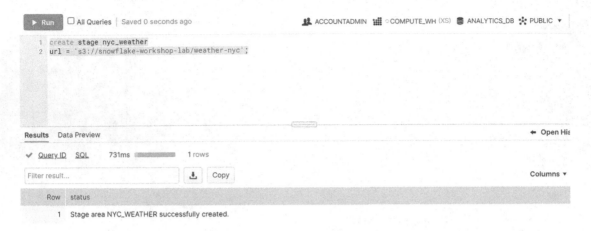

Figure 11-2. *Creating an External Stage Related to an AWS S3 Bucket*

This query creates a stage from where our weather sample unstructured data has been previously loaded into an AWS S3 bucket. Now let's take a closer look at the contents of the newly created nyc_weather stage by executing Snowflake's LIST command in Figure 11-3.

```
list @nyc_weather;
```

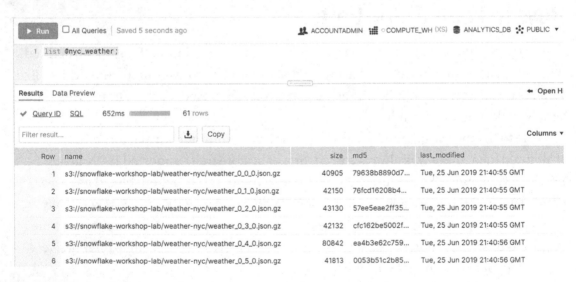

Figure 11-3. *Contents of Stage*

The preceding output shows many JSON files that are stored on the AWS S3 bucket. Now let us run our next SQL statement as in Figure 11-4 to load the data through Snowflake's standard loading command COPY INTO:

```
copy into json_weather_data
from @nyc_weather
file_format = (type=json);
```

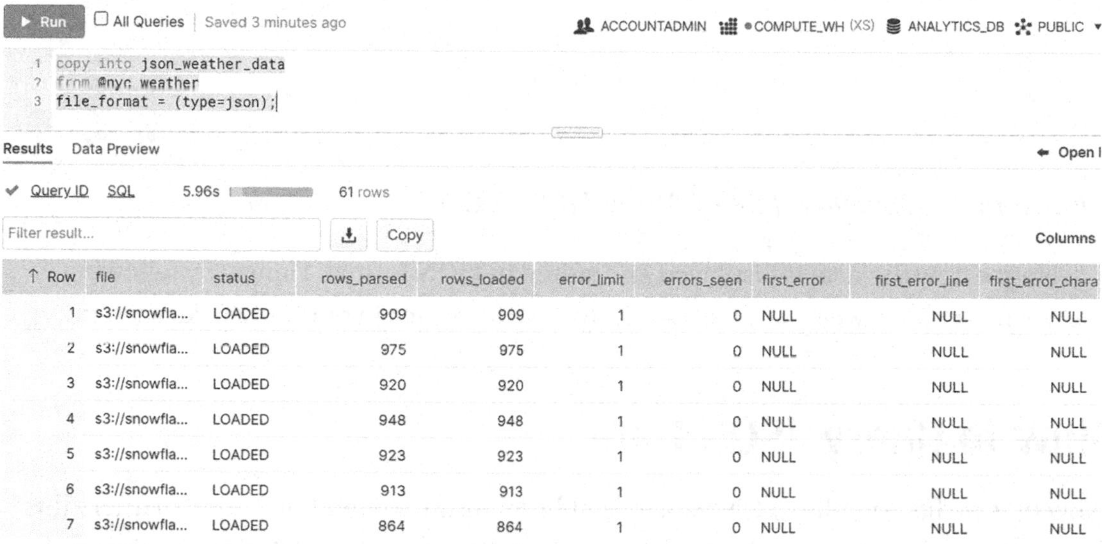

Figure 11-4. *Copy Into JSON_WEATHER_DATA*

Once executed, this command will load data from the AWS S3 bucket into the JSON_WEATHER_DATA table we created earlier. Here we are also specifying file_format as JSON as part of the COPY INTO command so Snowflake knows the File Format type and how to process the files when loading them.

Now let's check if data is loaded properly in the table. Use the SELECT command in Figure 11-5 to verify you have JSON data loaded into the JSON_WEATHER_DATA table.

```
select * from json_weather_data limit 5;
```

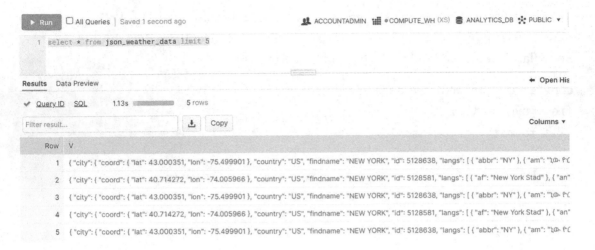

Figure 11-5. *Contents of JSON_WEATHER_DATA*

Notice how the data is displayed and stored in JSON format. Hence, by following these simple steps, we have successfully loaded JSON files in our Snowflake account.

How to Query JSON Data

Now that we have loaded JSON data in a table, let us understand how we can query this data. You can easily access elements in JSON objects using the JSON dot notation and bracket notation. **In both cases the column name is case-insensitive, but the element name is case-sensitive. Both enable to pull out values at lower levels in the JSON hierarchy allowing each field to be treated as a column in a relational table.** Here is an example of how to query a JSON object:

Dot notation syntax:

```
<column>:<level1_element>.<level2_element>.<level3_element>
```

Bracket notation syntax:

```
<column>['<level1_element>']['<level2_element>']
```

In our example in Figure 11-6, we use dot notation to query JSON data using the following command:

```
select distinct v:city.country from json_weather_data where
v:city.id=5128638
```

Figure 11-6. *Query JSON_WEATHER_DATA.*

Tip   The preceding example gives you the value of country for a particular city_id. Here v refers to the column in the json_weather_data table, and city.country is an element in the JSON schema. Snowflake allows you to effectively specify a column within the column (i.e., a sub-column), which is dynamically derived based on the schema definition embedded in the JSON data. This notation is very similar to traditional SQL language, which makes this very easy to learn.

How to Create a View on JSON Data

In traditional SQL concepts, views are defined as virtual tables. They help in combining, segregating, and protecting data. Now we can create views on top of data loaded in Snowflake including JSON data within a VARIANT column. This enables users to see data in a very clean and organized manner. In Figure 11-7 we show you how easy it is to create a view on top of the JSON_WEATHER_DATA table with the JSON data within it.

```
create view weather_data_view as
select
v:time::timestamp as time,
v:city.id::int as city_id,
v:city.name::string as city_name,
v:city.country::string as country,
v:wind.speed as speed
from json_weather_data where v:city.id=5128638
```

Figure 11-7. *Create View WEATHER_DATA_VIEW.*

The preceding example creates a view named weather_data_view on top of the JSON_WEATHER_DATA table for New York City. This view has five fields and can be queried like a normal relational view (without any dot or bracket annotation now). Figure 11-8 shows how easy it is to SELECT data from this newly created relational view.

```
select distinct city_name,country from weather_data_view;
```

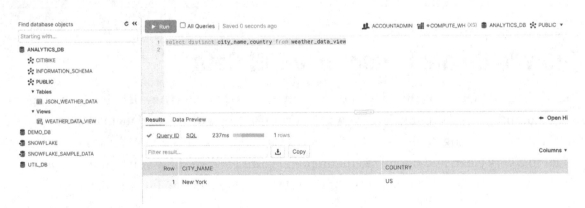

Figure 11-8. *Query View WEATHER_DATA_VIEW.*

Please note that on the left of the UI in database objects, you can see the new relational view WEATHER_DATA_VIEW that we just created.

How to Perform a Join Operation on a View

In the previous section, we learned how we can create a relational view on JSON semi-structured data. Now once the view is created, we can consider it as a simple relational view and perform standard join operations on top of it. We will again demonstrate this using our sample "ANALYTICS_DB"."CITIBIKE"."CITIBIKE_NYC" table in our sample CITIBIKE schema and ANALYTICS_DB that we create and load in Chapter 12.

Figure 11-9 shows how to execute a SELECT query on the CITIBIKE_NYC table.

```
select * from citibike.citibike_nyc limit 5;
```

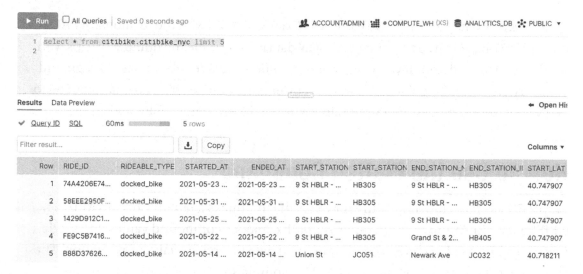

Figure 11-9. *Query View CITIBIKE_NYC.*

CITIBIKE_NYC is a table that contains Citi Bike data from New York City. We will use this table to join with our WEATHER_DATA_VIEW using the query as shown in Figure 11-10.

```
select count(*) as num_trips
from citibike.citibike_nyc
left outer join weather_data_view
on date_trunc('hour',time)=date_trunc('hour',started_at)
and speed>7;
```

Figure 11-10. Join Between CITIBIKE_NYC and WEATHER_DATA_VIEW

In the preceding example, we easily join CITIBIKE_NYC with the relational WEATHER_
DATA_VIEW based on semi-structured JSON data to determine the number of trips with
wind_speed more than 7. If you come from a traditional relational database background,
this is extremely exciting since it shows the power of Snowflake being able to allow you to
easily analyze and process JSON data alongside regular relational tables and views.

How to Use the Flatten Command

FLATTEN is a table function with Snowflake that produces a lateral view of VARIANT,
OBJECT, or ARRAY columns. In other words, it explodes compound values into single
rows. It can be used to convert semi-structured data to a relational representation. The
following is the syntax used for the FLATTEN command:

```
FLATTEN(INPUT=><expr>[,PATH=><constant_expr>]
[,OUTER=>TRUE|FALSE]
[,RECURSIVE=>TRUE|FALSE]
[,MODE=>'OBJECT'|'ARRAY'|'BOTH'])
```

Here INPUT is a required parameter, and the other parameters are optional. Now for
us to understand how the FLATTEN command will work, let's first understand the data
in the JSON_WEATHER_DATA table with an example SELECT of it using the command
as shown in Figure 11-11.

```
select * from
json_weather_data
limit 5;
```

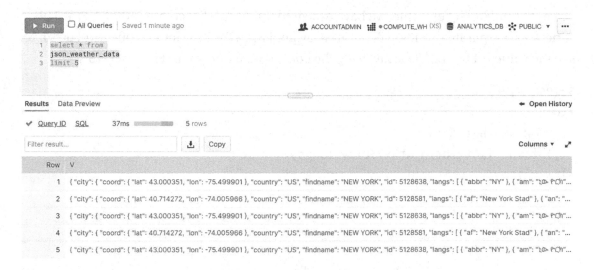

Figure 11-11. *Contents of JSON_WEATHER_DATA*

The preceding table are the contents of JSON_WEATHER_DATA. Now let's run the FLATTEN command and understand how it works using the command as shown Figure 11-12.

```
select v,value
from
  json_weather_data
 ,lateral flatten(input=>v:city)
 limit 20;
```

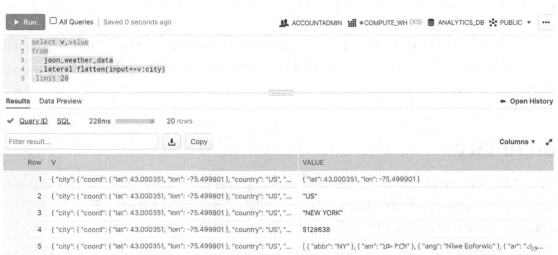

Figure 11-12. *Result of Applying FLATTEN Command on City*

In the preceding example, we have executed the FLATTEN command on CITY. This generates a separate row for every non-array field in CITY. Now let's go one level down to see the values of the LANG array using the command as shown in Figure 11-13.

```
select v,value
from
   json_weather_data
   ,lateral flatten(input=>v:city.langs)
limit 20
```

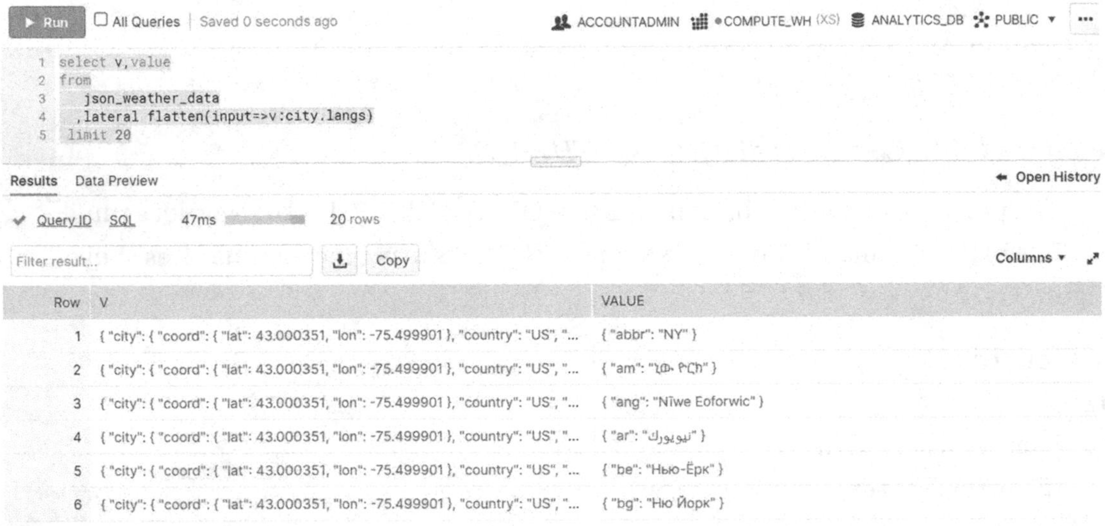

Figure 11-13.   *Result of Applying FLATTEN Command on LANG*

As we can see, the command generates separate rows for every value from the LANG array. In this way we can use the FLATTEN command and create a VIEW and use the same as standard SQL queries.

Summary

Snowflake supports semi-structured data enabling users to manage files like JSON and XML along with relational data. With Snowflake, users can choose to FLATTEN nested objects into a relational table or store them in native format within the VARIANT data type. This is simpler, scalable, and cost-effective and brings the true value of data democratization. The entire process is highly accessible and optimized behind the scenes allowing users to create simple views and SQL queries to extract data. Available as a service on leading cloud providers, Snowflake eliminates the architectural complexity helping customers integrate a wide range of data sources giving opportunities for data monetization.

CHAPTER 12

Loading Data

To take advantage of the cloud-native features, organizations are moving to the cloud and building their modern data platforms around Snowflake to support their ever-growing demand for data and analytical workloads.

Snowflake supports various workloads for data engineering, data lake, data warehouse, data science, data applications, and data sharing on their advanced Snowflake cloud data platform (CDP).

To build all these capabilities, data is the key, and it needs to be loaded into Snowflake. There are many ways to load data into Snowflake, including various data integration tools available from third-party vendors.

In this chapter, we will discuss loading data into Snowflake through their internal tools including their web interface and the SnowSQL command-line interface (CLI).

We are going to discuss database objects that are typically used during data loading in Snowflake. If you are coming from a traditional RDBMS background, these may be new to you. The new database objects we are going to introduce are stages (external and internal) and File Formats.

We will use the Citi Bike NYC dataset to demonstrate various ways to load data into Snowflake.

Let's dig in!

Snowflake Object Hierarchy

Before we explore data load options, let's understand the objects and the hierarchy. In Figure 12-1 are some of the objects and the hierarchy.

© Frank Bell, Raj Chirumamilla, Bhaskar B. Joshi, Bjorn Lindstrom, Ruchi Soni, Sameer Videkar 2022
F. Bell et al., *Snowflake Essentials*, https://doi.org/10.1007/978-1-4842-7316-6_12

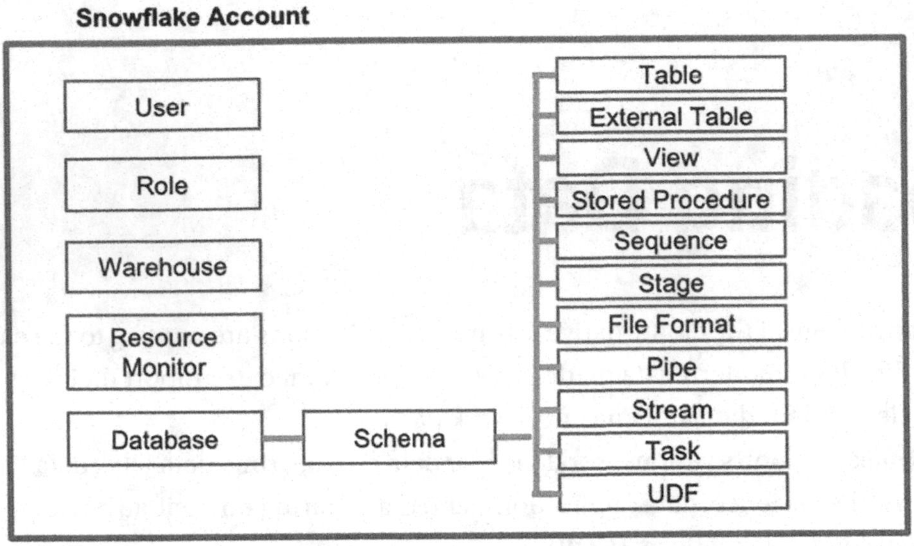

Snowflake Account

Figure 12-1. Object Hierarchy in SnowflakeCredits: Snowflake

Object types in the first column on the left are high-level objects or container objects, as you see only Database has underlying schema(s) objects, that is, you can create multiple schemas under a database. And further, all the objects in the third column on the right can be created within a schema object, and these objects are grouped as securable objects.

Let's explore some of the objects that we need to create before loading the data in.

Stage

A **stage** is a path to files on a cloud provider's storage, for example, a path to a file in an AWS S3 bucket or Azure Blob storage or Google Cloud Storage.

Snowflake extends this functionally and provides support for External Stages and Internal Stages.

An **External Stage** is a file location on a cloud provider's storage that can be accessed by creating a *Named External Stage* object in Snowflake. Managing access permissions to these locations is done through cloud IAM roles and/or Access Control Lists (ACLs).

Internal Stage(s) allow files to be stored within and managed by Snowflake.

There are three types of Internal Stages:

> User Stages: Whenever a user is created in Snowflake, a default stage is created and allocated to store the files by the user. User stages are referenced using the @~ notation. The files in this stage are only accessible by the user, and this type of stage cannot be dropped or altered.

> Table Stages: Whenever a table is created in Snowflake, a default stage is created and allocated to store files for that table. Table stages have the same name as the actual table. Table stages are referenced using the *@%table_name* notation. The files in the table stage are accessible to users who have privileges to read from the table.

> Named Stages: Internal Named Stage objects point to files that are stored in the location that is managed by Snowflake. These files are not accessible directly from outside Snowflake and can't be managed through cloud IAM roles or ACLs. These files can only be accessed from within Snowflake, and Snowflake's Access Control Framework can be used to manage access permissions.

File Format

A **File Format** is a predefined structural definition of a data file that can be defined with a variety of parameters like compression, file type, delimiter, data/time formats, encoding, etc.

File Formats are independent objects and can be reused with COPY commands when loading similar structured datasets.

File Formats can also be managed within Role-Based Access Control Frameworks.

We will be working with Internal Named Stages and File Formats later in this chapter.

Public Dataset

Before we start creating database objects and loading data, let's first download the Citi Bike NYC dataset. This is a public dataset available for analysis, development, and visualizations. Data use policy can be found at `www.citibikenyc.com/data-sharing-policy`.

Download the Dataset

Go to the URL `https://ride.citibikenyc.com/system-data` [updated url].

This is the main website with all of this wonderful Citi Bike data. For our example, click Download Citi Bike trip history data.

This link will take you to the AWS website where the Citi Bike historical data files reside which currently is

`https://s3.amazonaws.com/tripdata/index.html`

This is technically the open AWS bucket that has been created to share the Citi Bike data. Let's download a zip file from 2021 at this link here:

`https://s3.amazonaws.com/tripdata/202109-citibike-tripdata.csv.zip`

Save the zip file on to your desktop and extract the csv file.

The csv file has data for the following attributes:

```
RIDE_ID          STRING
RIDEABLE_TYPE      STRING
STARTED_AT TIMESTAMP
ENDED_AT    TIMESTAMP
START_STATION_NAME      STRING
START_STATION_ID STRING
END_STATION_NAME         STRING
END_STATION_ID    STRING
START_LAT    DECIMAL(8,6)
START_LNG    DECIMAL(9,6)
END_LAT        DECIMAL(8,6)
END_LNG      DECIMAL(9,6)
MEMBER_CASUAL    STRING
```

We will create a table with the above-mentioned definition in the database and load the downloaded csv file into that table.

Loading Data into Snowflake via Web UI

If you don't have a Snowflake account, you can request a trial account. Snowflake made it very easy to request a trial account from the URL https://signup.snowflake.com/.

If you already have a Snowflake user account, then you can follow the steps to load the csv file that was downloaded earlier.

Log In to Your Snowflake Account

Log into your account with the Snowflake-provided URL and user ID and password that you created during the initial sign-up. Figure 12-2 shows the sign-in page for a Snowflake account.

Figure 12-2. *Snowflake Sign-In Page*

Create a Database

To create a table and load the data, we need to create a container object, that is, a database, first. To create a database, click the 🗄 Databases icon near the top-left corner of the Snowflake landing page. And then click ⊕ Create…, which prompts for database details. Figure 12-3 shows the list of databases within the account.

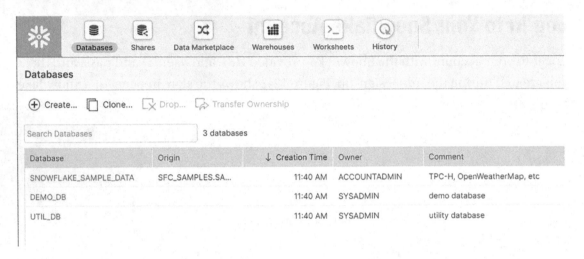

Figure 12-3. *Databases Page*

On the Create Database pop-up screen, enter value ANALYTICS_DB for the Name parameter (required), populate the value for the Comment parameter (optional), and then click Finish .

Figure 12-4 shows the pop-up screen to enter database name and comment.

Create Database

Name *	ANALYTICS_DB
Comment	Database for Citibike Data

Show SQL   Cancel   Finish

Figure 12-4. *Create Database Pop-Up Screen*

Create a Schema

Once the database is created, click the ANALYTICS_DB database name. Figure 12-5 shows the Databases page with ANALYTICS_DB listed.

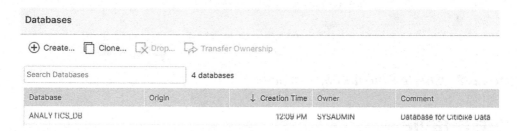

Figure 12-5. *Databases Page with a List of Existing Databases*

Now it's time to create a schema within the ANALYTICS_DB database. Go to the Schemas tab and click ⊕ Create… . Figure 12-6 shows the list of schemas within the ANALYTICS_DB database.

Databases › **ANALYTICS_DB**

Tables	Views	**Schemas**	Stages	File Formats	Sequences	Pipes

⊕ Create…   Clone…   Alter…   Drop…   Transfer Ownership

Schema	Creation Ti...	Owner	Managed Access	Comment
INFORMATION_SCHEMA	2:01:07 PM			Views describing the contents of schemas in this database
PUBLIC	1:55:34 PM	SYSADMIN		

Figure 12-6. *Page with a List of Existing Schemas Within the ANALYTICS_DB Database*

To complete the schema creation step, in the pop-up screen for Create Schema, enter value CITIBIKE for the Name parameter (required), populate the value for the Comment parameter (optional), and then click Finish .

Figure 12-7 shows the pop-up screen to enter schema name and comment.

Create Schema

Name*	CITIBIKE
Comment	Schema for Citibike Data|

☐ Managed Access

Show SQL   Cancel   Finish

Figure 12-7. *Create Schema Pop-Up Screen*

Create a Table

Now the database and schema are created. In the next steps, we will create a table and load the Citi Bike NYC data into that table.

Let us create a table in the CITIBIKE schema within the ANALYTICS_DB database.

Click the Databases icon 🛢 Databases near the top-left corner.

Figure 12-8 shows the Databases page with ANALYTICS_DB listed.

Figure 12-8. *Databases Page with Database Listing*

From the list of databases, click the ANALYTICS_DB database name and then select Tables tab. Figure 12-9 shows the list of tables within ANALYTICS_DB (at this point it's empty).

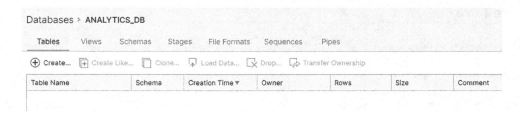

Figure 12-9. *Tables Within the ANALYTICS_DB database*

To create a new table, click the ⊕ Create... icon and in the pop-up screen enter values for the parameters. Table Name is required, for Schema Name the default CITIBIKE value should have been selected in the dropdown (if not, pick the value manually from the dropdown), enter the value for Comment, and add columns by clicking the ⊕ Add icon in the column input area (lists of column names and data types have been provided earlier in the section "Download the Dataset"). You must hit Add for each column that you are going to add to the table. After entering all the columns, click Finish.

For your convenience, a DDL script to create the CITIBIKE_NYC table has been provided at the end of this section. You can run the script in a worksheet and may skip this step to create a table through a wizard.

Figure 12-10 shows all the parameters that need to be populated to create a table.

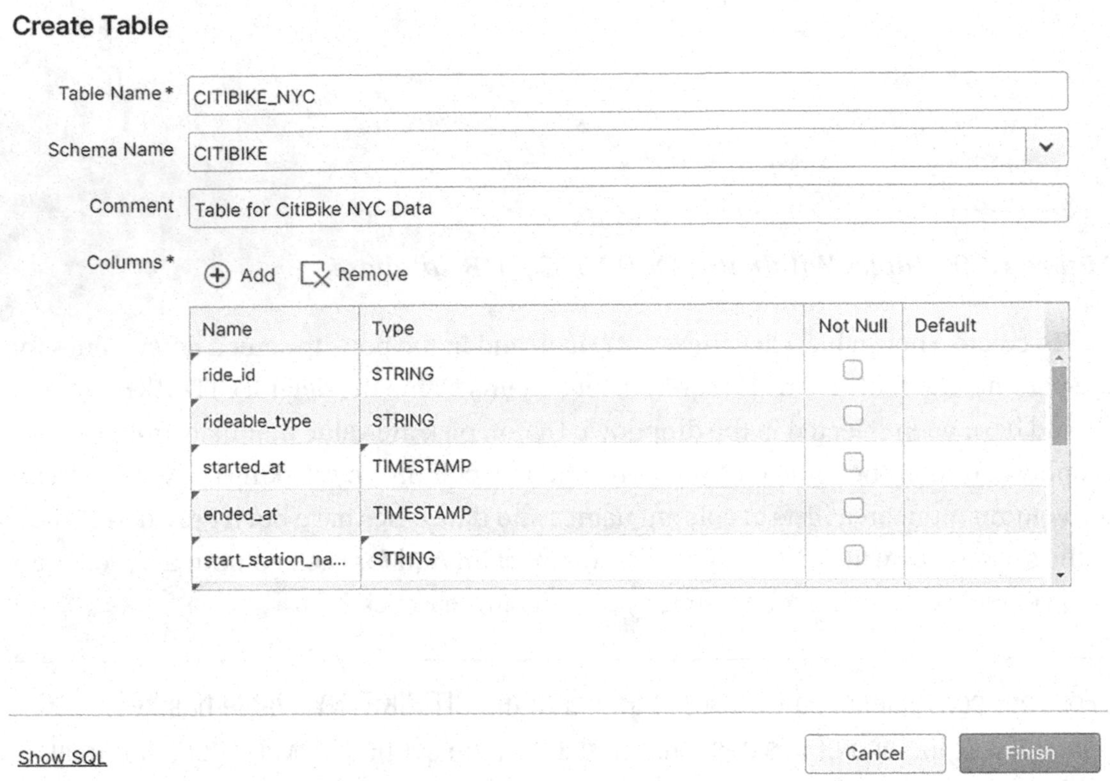

Create Table

Table Name *   CITIBIKE_NYC

Schema Name   CITIBIKE   ∨

Comment   Table for CitiBike NYC Data

Columns *   ⊕ Add   ⊏ₓ Remove

Name	Type	Not Null	Default
ride_id	STRING	☐	
rideable_type	STRING	☐	
started_at	TIMESTAMP	☐	
ended_at	TIMESTAMP	☐	
start_station_na...	STRING	☐	

Show SQL   Cancel   Finish

Figure 12-10. *Create Table Pop-Up Screen*

Once the table has been created successfully, you should see the table definition like the following screen. Figure 12-11 shows the definition of the CITIBIKE_NYC table in the CITIBIKE schema within the ANALYTICS_DB database.

Databases > ANALYTICS_DB > **CITIBIKE_NYC (CITIBIKE)**

Tables   Views   Schemas   Stages   File Formats   Sequences   Pipes

⬆ Load Table

Column Name	Ordinal ▲	Type	Nullable	Default	Comment
RIDE_ID	1	VARCHAR(16777216)	true	NULL	
RIDEABLE_TYPE	2	VARCHAR(16777216)	true	NULL	
STARTED_AT	3	TIMESTAMP_NTZ(9)	true	NULL	
ENDED_AT	4	TIMESTAMP_NTZ(9)	true	NULL	
START_STATION_NAME	5	VARCHAR(16777216)	true	NULL	
START_STATION_ID	6	VARCHAR(16777216)	true	NULL	
END_STATION_NAME	7	VARCHAR(16777216)	true	NULL	
END_STATION_ID	8	VARCHAR(16777216)	true	NULL	
START_LAT	9	VARCHAR(16777216)	true	NULL	
START_LNG	10	VARCHAR(16777216)	true	NULL	
END_LAT	11	VARCHAR(16777216)	true	NULL	
END_LNG	12	VARCHAR(16777216)	true	NULL	
MEMBER_CASUAL	13	VARCHAR(16777216)	true	NULL	

Figure 12-11. *List of Columns for Table CITIBIKE_NYC table*

Instead of creating the table through a wizard, you can run the following DDL in a worksheet. It's an easy alternative approach, but you will get the same result:

```
CREATE TABLE "ANALYTICS_DB"."CITIBIKE"."CITIBIKE_NYC"
("RIDE_ID" STRING,
 "RIDEABLE_TYPE" STRING,
 "STARTED_AT" TIMESTAMP,
 "ENDED_AT" TIMESTAMP,
 "START_STATION_NAME" STRING,
 "START_STATION_ID" STRING,
 "END_STATION_NAME" STRING, "END_STATION_ID" STRING,
 "START_LAT" DECIMAL(8,6),
 "START_LNG" DECIMAL(9,6),
```

```
"END_LAT" DECIMAL(8,6),
"END_LNG" DECIMAL(9,6),
"MEMBER_CASUAL" STRING)
COMMENT = 'Table for CitiBike NYC Data';
```

Load Table

To load data from the csv file that we downloaded earlier into the table, click ⬆ Load Table
right below the Tables tab.

Select a Warehouse

Since this is a load operation, we need to provide a warehouse (compute) in the load
context. Select a warehouse from the dropdown and click [Next].

Figure 12-12 shows a pop-up wizard to load data from a source file into the table.

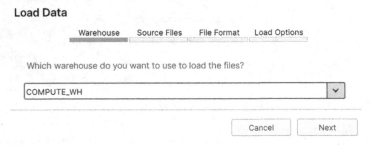

Figure 12-12. *Pick a Warehouse.*

Select Source File(s)

Keep the "Load files from your computer" option selected and click
the [Select Files...] icon.

Figure 12-13 shows source file selection options.

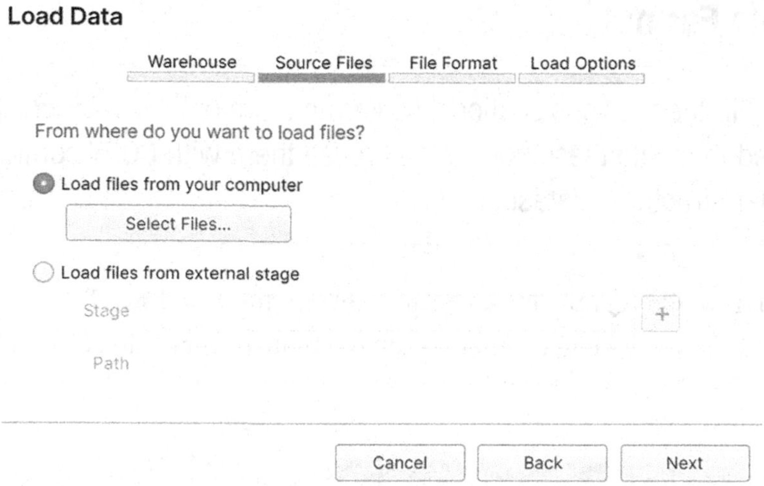

Load Data

Warehouse   Source Files   File Format   Load Options

From where do you want to load files?

◉ Load files from your computer

Select Files...

◯ Load files from external stage

Stage   +

Path

Cancel   Back   Next

Figure 12-13. *Select Source Files to Be Loaded*

From the actual list of files, select the extracted csv file that was downloaded earlier and click Next .

Figure 12-14 shows the selected source file to be loaded.

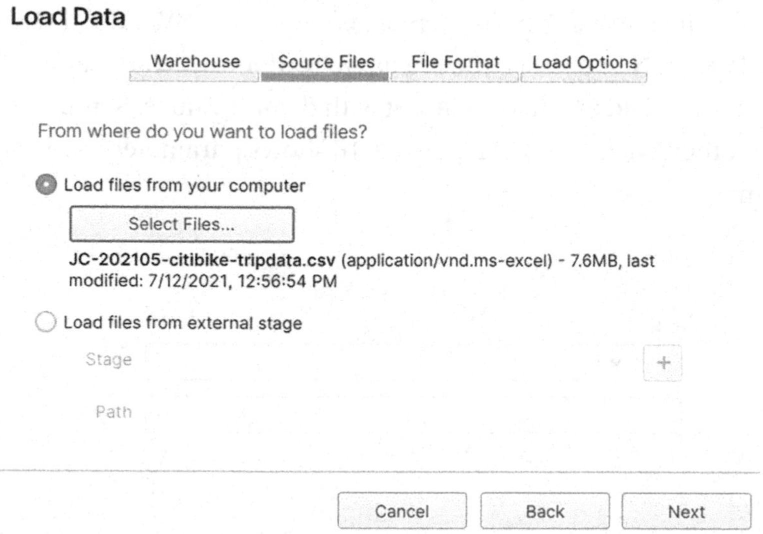

Load Data

Warehouse   Source Files   File Format   Load Options

From where do you want to load files?

◉ Load files from your computer

Select Files...

JC-202105-citibike-tripdata.csv (application/vnd.ms-excel) - 7.6MB, last modified: 7/12/2021, 12:56:54 PM

◯ Load files from external stage

Stage   +

Path

Cancel   Back   Next

Figure 12-14. *Selected File to Be Loaded with File Properties*

Create a File Format

As mentioned in the previous section, File Formats are optional objects. But it is recommended to create these objects and reuse them with COPY commands when loading similar structured datasets.

To create a new File Format, click ⊞ right next to the dropdown list.
Figure 12-15 shows the File Format screen to create or pick File Formats.

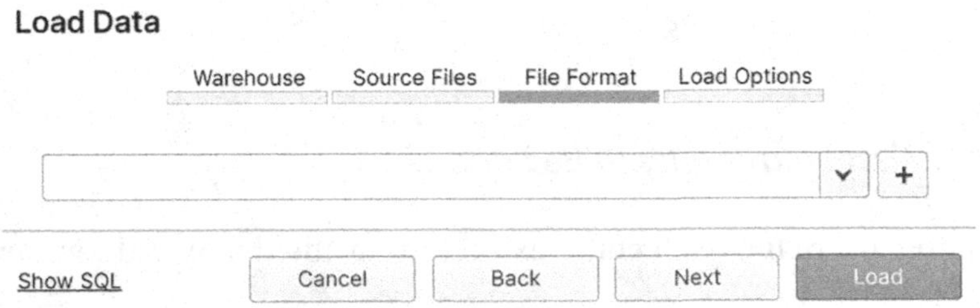

Load Data

Warehouse	Source Files	File Format	Load Options

	∨	+

Show SQL   Cancel   Back   Next   Load

Figure 12-15. *File Format Selection*

In the "Create File Format" pop-up screen, enter value CSV_FILE_FORMAT for parameter File Format Name. Since the file has a header row, enter value 1 for the "Header lines to skip" field and leave the rest with default values. Scroll down to review more details and then click Finish. Figure 12-16 shows parameters and values to create a new File Format.

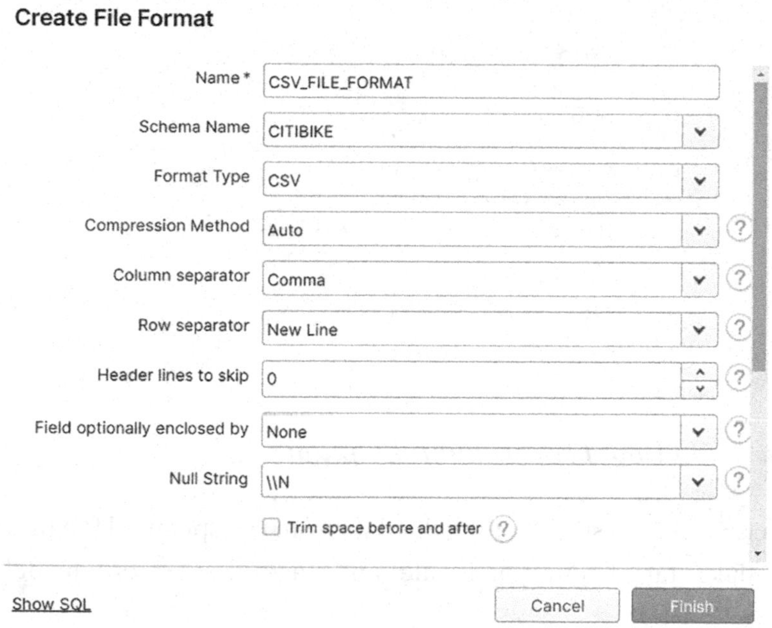

Create File Format

Name *	CSV_FILE_FORMAT
Schema Name	CITIBIKE ⌄
Format Type	CSV ⌄
Compression Method	Auto ⌄ (?)
Column separator	Comma ⌄ (?)
Row separator	New Line ⌄ (?)
Header lines to skip	0 ⌃⌄ (?)
Field optionally enclosed by	None ⌄ (?)
Null String	\\N ⌄ (?)
	☐ Trim space before and after (?)

Show SQL   Cancel   **Finish**

Figure 12-16. *Create File Format Pop-Up Screen*

As the newly created File Format (CSV_FILE_FORMAT) is populated now from the dropdown, click [Next]. Figure 12-17 shows the selected File Format from the dropdown.

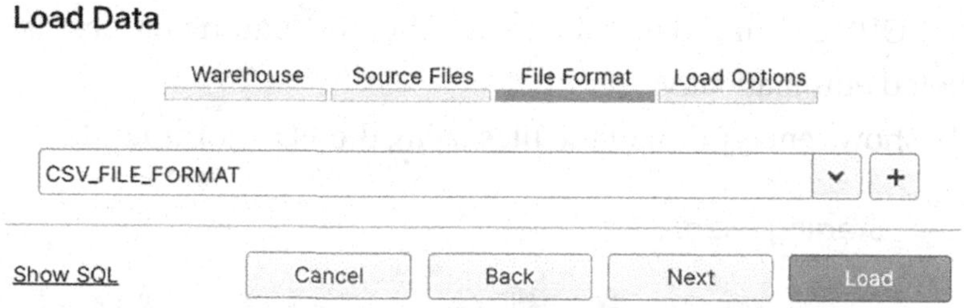

Load Data

Warehouse	Source Files	File Format	Load Options

CSV_FILE_FORMAT ⌄ +

Show SQL   Cancel   Back   Next   **Load**

Figure 12-17. *File Format Selection from the Dropdown Menu*

Error Handling

The final step in the load data process is to specify an error handling option. As you can see, there are four different options to select from. The options' descriptions are self-explanatory, and let's keep the default option selected.

Figure 12-18 shows load options for error handing during data load.

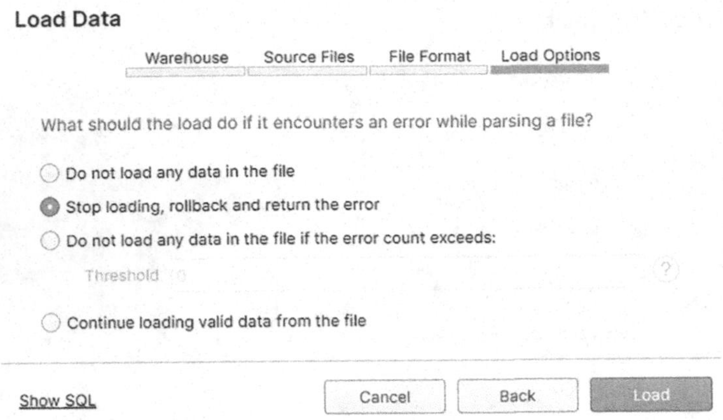
Load Data
Warehouse
Source Files
File Format
Load Options
What should the load do if it encounters an error while parsing a file?
Do not load any data in the file
Stop loading, rollback and return the error
Do not load any data in the file if the error count exceeds:
Threshold
Continue loading valid data from the file
Show SQL
Cancel
Back
Load

***Figure 12-18.** Load Data Error Handling Options*

Finally, click 
Load
to start loading the data from the specified file into the table.

Now, Snowflake starts to encrypt the files and stages them internally before loading into the table.

During the data loading process, PUT and COPY commands are used behind the scenes. We will discuss PUT and COPY commands later in the section.

The Snowflake-managed stage is used to temporarily stage the file(s) before it gets loaded with the COPY command into the table. After the loading process is successful, the file gets deleted automatically.

Figure 12-19 shows encryption of the files using the PUT command.

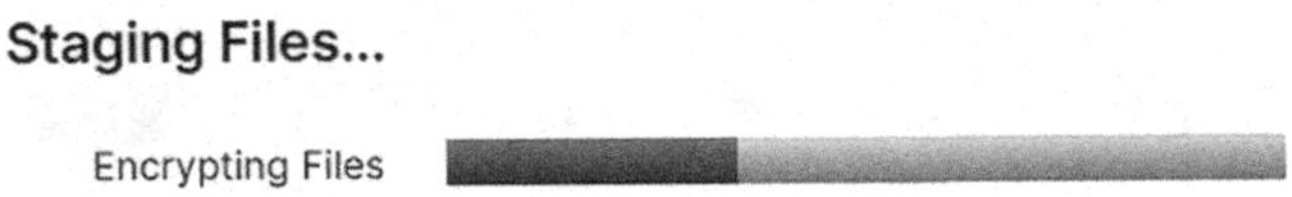
Staging Files...
Encrypting Files

***Figure 12-19.** File Encryption Before Moving Files onto the Stage*

Figure 12-20 shows staging of the encrypted files using the PUT command.

Staging Files...
Encrypted Files
Staging Files

***Figure 12-20.** Moving Encrypted Files to Stage*

When the load is successful, you should see the results screen like Figure 12-21. The file name may be different (depending on the latest file you have downloaded). Figure 12-21 shows the Load Results screen that reflects rows parsed and rows successfully loaded.

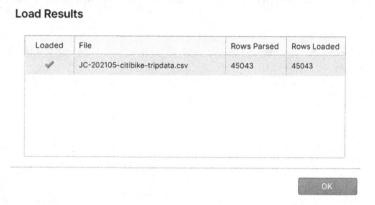

Load Results

Loaded	File	Rows Parsed	Rows Loaded
✓	JC-202105-citibike-tripdata.csv	45043	45043

OK

Figure 12-21. *Results Screen with Parsed vs. Loaded Rows from the File*

Verify the Results

To check the number of rows loaded and the size of the table, click [>_ Worksheets] at the top and then expand the ANALYTICS_DB database and CITIBIKE schema and hover the mouse over to CITIBIKE_NYC table. You should see the details like the following. Figure 12-22 shows the number of rows and size of the table that was loaded.

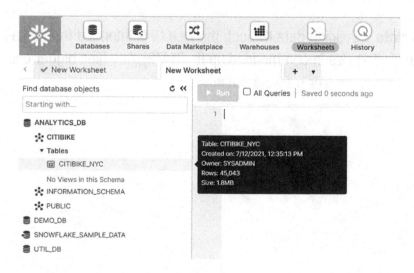

Figure 12-22. *Table Metadata with Name, Created on, Owner, Rows, and Size*

If you would like to see the rows from the table, run a select statement on the table in a worksheet:

```
select * from ANALYTICS_DB.CITIBIKE.CITIBIKE_NYC;
```

Figure 12-23 shows the results from the table when queried from a worksheet.

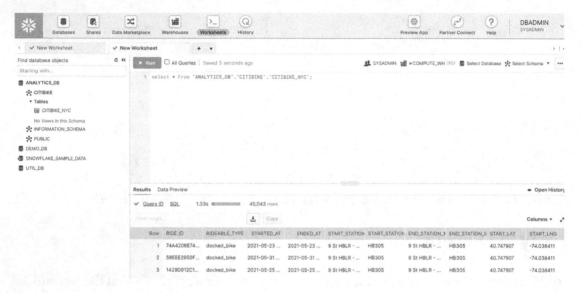

Figure 12-23. *Query and the Results*

This completes loading data through the web UI into a table.

Important Note   Loading data through the web UI is intended for a small number of files with limited size (recommended up to 50 MB). Any files bigger could lead to performance issues and/or browser crash.

Drop and Recreate Objects

Drop Objects

We will drop objects that were created through web UI in the previous section.

Open a new worksheet and run the following commands:

```
use role sysadmin;
-- The following command drops the database
-- and all the objects within.
drop database analytics_db;
```

Dropping the database will drop all the objects within. There is no need to individually drop the objects, for example, schema, File Format, table, etc.

Recreate Objects

Now we will recreate the objects that are needed to load the data again through SnowSQL. We will create these objects through SQL commands.

To do this, open a new worksheet and run the following commands:

```
use role sysadmin;
--Create a database
create database analytics_db comment ="Database for Citibike Data";
create schema  citibike comment="Schema for Citibike Data";
use database analytics_db;
use schema citibike;

--Create a table
CREATE TABLE "ANALYTICS_DB"."CITIBIKE"."CITIBIKE_NYC"
("RIDE_ID" STRING,
 "RIDEABLE_TYPE" STRING,
 "STARTED_AT" TIMESTAMP,
 "ENDED_AT" TIMESTAMP,
 "START_STATION_NAME" STRING,
 "START_STATION_ID" STRING,
 "END_STATION_NAME" STRING, "END_STATION_ID" STRING,
```

```
 "START_LAT" DECIMAL(8,6),
 "START_LNG" DECIMAL(9,6),
 "END_LAT" DECIMAL(8,6),
 "END_LNG" DECIMAL(9,6), "MEMBER_CASUAL" STRING)
 COMMENT = 'Table for CitiBike NYC Data';

--Create a File Format
 CREATE FILE FORMAT "ANALYTICS_DB"."CITIBIKE".CSV_FILE_FORMAT
 TYPE = 'CSV'
 COMPRESSION = 'AUTO'
 FIELD_DELIMITER = ','
 RECORD_DELIMITER = '\n'
 SKIP_HEADER = 1
 FIELD_OPTIONALLY_ENCLOSED_BY = 'NONE'
 TRIM_SPACE = FALSE
 ERROR_ON_COLUMN_COUNT_MISMATCH = TRUE
 ESCAPE = 'NONE'
 ESCAPE_UNENCLOSED_FIELD = '\134'
 DATE_FORMAT = 'AUTO'
 TIMESTAMP_FORMAT = 'AUTO'
 NULL_IF = ('\\N');

--Create a Named Internal Stage
 CREATE STAGE "ANALYTICS_DB"."CITIBIKE".SNFK_INTERNAL_STAGE
 COMMENT = 'Snowflake Managed Named Internal Stage';
```

If you recall, when loading data through the web UI, Snowflake uses an internal staging area to move the files, loads them into a table, and deletes them when the load is complete.

But, in this section, we are going to use the Named Internal Stage (SNFK_INTERNAL_STAGE) to move the files with the PUT command and then use the COPY command to load into a table.

PUT and COPY Commands

PUT Command

This supports uploading files from local storage directly on a client machine to Snowflake's Internal Stages, that is, Table, User, and Named Stages. The differences between these stages are explained earlier in this chapter.

Snowflake does not support uploading files to External Stages with the PUT command.

The PUT command is not supported to run explicitly in the web UI, but it is supported with ODBC and JDBC drivers and Python Connector. It always encrypts and compresses files before copying into an Internal Stage.

The PARALLEL parameter can be specified for the number of threads to use for uploading the files to Internal Stages.

COPY into <table> Command

COPY is a SQL command, which is used to (bulk) load data from file(s) into table(s) with parallelism.

With the COPY command, files can be loaded from external locations or Named External Stages that reference external locations (Amazon S3, Google Cloud Storage, or Microsoft Azure).

When files are staged into an Internal Stage with the PUT command, the COPY command can be used to load data from staged files into tables.

Copy supports loading data from File Formats of type CSV, JSON, AVRO, ORC, PARQUET, or XML. Individual file names or patterns with regular expressions can be mentioned to identify the files.

The VALIDATION_MODE parameter helps to identify errors or issues with the files without loading them into tables.

The following validation options specify the return results:

- RETURN_<n>_ROWS: Validates <n> number of rows,

- for example, RETURN_10_ROWS validates the first ten rows

- RETURN_ERRORS: Returns all errors across all files

- RETURN_ALL_ERRORS: Returns all errors across all files including from files that were partially loaded earlier

273

The **VALIDATION_MODE** parameter is not supported with COPY statements that transform data during loading.

The ON_ERROR parameter can be used to specify an action to perform when there are errors during the data load.

The following values specify the actual action:

- CONTINUE: Continue to load the files into a table.

- SKIP_FILE: Skips any file if there are any errors.

- SKIP_FILE_<num>: Skips any file if the number of errors equals or exceeds the number specified in the parameter, that is, SKIP_FILE_10 (ten-error threshold).

- SKIP_FILE_<num>%: Skips any file if the number of errors equals or exceeds the percentage specified in the parameter, that is, SKIP_FILE_10% (10% error threshold).

- ABORT_STATEMENT: Aborts the operation when any errors are encountered during the load. This is the default value.

Some transformation functions are supported with the COPY command during the data load (refer to documentation for more details).

Loading Data via SnowSQL

SnowSQL is a command-line interface (CLI) client developed using the Snowflake Connector for Python, and the CLI supports interactive and batch mode to connect to Snowflake's cloud data platform and run DDL and DML commands.

It is supported on various flavors of Linux, MacOS, and Windows versions.

Download SnowSQL

The CLI client is an executable and can be directly downloaded from Snowflake's account that you are logged into. To download the executable, click "Help" and then select "Download" from the menu.

Figure 12-24 shows where to go and download the SnowSQL CLI client.

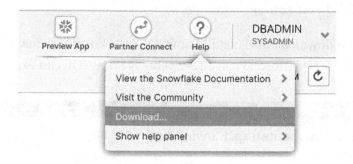

Figure 12-24. *Download SnowSQL CLI Client*

In the Downloads pop-up screen, download the latest version by clicking 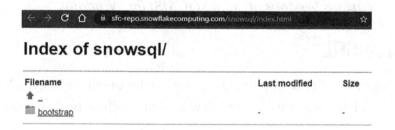. This will take you to the official Snowflake repo website.

Figure 12-25 shows Snowflake's public repository where the SnowSQL executables are located.

Figure 12-25. *Repository Home for SnowSQL*

Navigate through the folder structure and click the folder for the operating system on which you would like to install SnowSQL. Figure 12-26 shows the OS-level folder structure for the SnowSQL CLI executable.

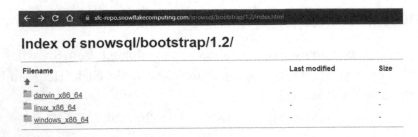

Figure 12-26. *OS-Level Folder Structure*

We are going to download and install SnowSQL on a Windows machine, so pick the latest modified MSI for Windows. Figure 12-27 shows SnowSQL CLI executables.

Index of snowsql/bootstrap/1.2/windows_x86_64/

Filename	Last modified	Size	SHA256
↑ ..	-	-	
snowsql-1.2.0-windows_x86_64.msi	2019-11-12T21:39:15	17428 kB	f446fa78611c606ed5a2fb82a9fc69128d3283a2af8b90332
snowsql-1.2.1-windows_x86_64.msi	2019-12-05T17:26:56	19189 kB	8f099bc9dd964f9c645fa53489265dac59ba298351be1abde
snowsql-1.2.10-windows_x86_64.msi	2020-09-14T19:10:32	44961 kB	5c86d20011fe51575aa939d2687fbcb9fa84fb49f8d6fd3e6
snowsql-1.2.11-windows_x86_64.msi	2021-02-10T03:23:53	45862 kB	277d9188b25d3e364a458c6592ce2cf59c55f9f08a9e371ae
snowsql-1.2.12-windows_x86_64.msi	2021-03-02T14:55:45	45776 kB	ea165446e24ebeb564f55592e58f877a9cafd4927e763f50c
snowsql-1.2.13-windows_x86_64.msi	2021-03-11T19:18:09	45973 kB	cb26863e4cce94a7cccb818584c87496abc05290bd5133b23
snowsql-1.2.14-windows_x86_64.msi	2021-03-18T22:55:07	45707 kB	4e585c426a85ec20fd4cba15796458e5863ef44a4962e0d99
snowsql-1.2.15-windows_x86_64.msi	2021-06-04T16:33:40	45793 kB	ca1f440207ea1c4f403aa91bff549bc05d84f4237bd0eb987
snowsql-1.2.16-windows_x86_64.msi	2021-06-21T23:58:21	45805 kB	8fedff4292aea7992b93bdeeae190b4303e3f0ed76fed9916

Figure 12-27. *Various Versions of SnowSQL MSI for Windows*

Install SnowSQL

Installing the SnowSQL CLI is very straightforward. After downloading the MSI for Windows, click it to open the installation wizard and follow the instructions on the screen.

Verify the installation by opening a command prompt and typing ***snowsql***. This should return help information. Figure 12-28 shows SnowSQL CLI options.

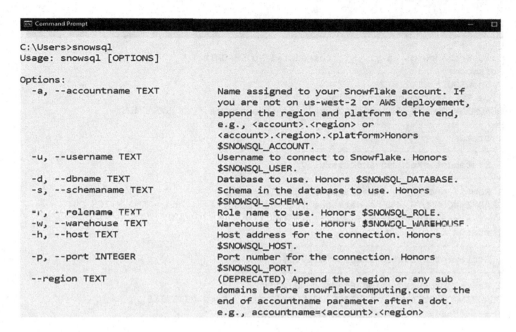

Figure 12-28. *Running snowsql cli from Command Prompt*

Log In and Set Context

To log in to Snowflake through SnowSQL, open a command prompt and run the `snowsql` cli command.

We are going to use -a (account name) and -u (username) parameters with SnowSQL. When prompted, enter the password:

e.g. > snowsql -a yxxxxxx.us-east-1 -u XXXXX

Account name could be different based on cloud provider and region. The account name is simply the snowflake account URL without the snowflakecomputing.com domain. This is how the Snowflake Account Name composed: account-name-itself. region-name.provider-name.

Once you log in to SnowSQL, set the context with user role, database, and schema. Figure 12-29 shows the snowsql command-line login information and context setting with use commands.

Figure 12-29. *Login with snowsql cli and Context Setting*

List Files from Internal Stage

List files from the Internal Stage that was created earlier. Results should show no files. Since we are using a Named Internal Stage, we need to prefix the name with "@". Figure 12-30 shows the command to list Named Internal Stage files from snowsql cli.

Figure 12-30. *File Listing from Named Internal Stage*

Stage Files with the PUT Command

Run a PUT command to copy local file(s) into the Internal Stage assuming the file is the *only* file in the C:\temp directory since we use the wildcard syntax of *.*.

For example, on Windows:

```
:> PUT file://c:\temp\*.* @SNFK_INTERNAL_STAGE
```

For example, on MacOS – assuming the file(s) are the only files in the /tmp directory:

```
:>PUT file:///tmp/*.* @SNFK_INTERNAL_STAGE
```

If you want to be specific to only one file, then you can do it like this on MacOS in the /tmp directory:

```
:>PUT file:///tmp/JC-202105-citibike-tripdata.csv @SNFK_INTERNAL_STAGE
```

If you want to be specific to only one file, then you can do it like this on MacOS in the /tmp directory:

```
:>PUT file://~/Downloads/JC-202105-citibike-tripdata.csv @SNFK_INTERNAL_STAGE
```

The PUT command will encrypt and compress the file before copying it into Internal Stage. Figure 12-31 shows the execution of PUT command to copy files from Temp directory on client machine into Name Internal Stage called SNFK_INTERNAL_STAGE.

```
DBADMIN#COMPUTE_WH@ANALYTICS_DB.CITIBIKE>put file://c:\temp\*.* @snfk_internal_stage;
JC-202105-citibike-tripdata.csv_c.gz(1.16MB): [##########] 100.00% Done (0.325s, 3.57MB/s).
+--------------------------------+--------------------------------+-------------+-------------+--------------------+--------------------+----------+---------+
| source                         | target                         | source_size | target_size | source_compression | target_compression | status   | message |
+--------------------------------+--------------------------------+-------------+-------------+--------------------+--------------------+----------+---------+
| JC-202105-citibike-tripdata.csv | JC-202105-citibike-tripdata.csv.gz | 7969936 |     1214729 | NONE               | GZIP               | UPLOADED |         |
+--------------------------------+--------------------------------+-------------+-------------+--------------------+--------------------+----------+---------+
1 Row(s) produced. Time Elapsed: 2.494s
DBADMIN#COMPUTE_WH@ANALYTICS_DB.CITIBIKE>
```

Figure 12-31. *PUT Command to Stage Files*

Run a LIST command to list the files in the Internal Stage:

```
e.g. :> LIST @SNFK_INTERNAL_STAGE
```

Figure 12-32 shows a list of files from Named Internal Stage SNFK_INTERNAL_STAGE. As you can see, the file has been compressed with gzip automatically.

```
DBADMIN#COMPUTE_WH@ANALYTICS_DB.CITIBIKE>list @snfk_internal_stage;
+------------------------------------------------------+---------+----------------------------------+-------------------------------+
| name                                                 | size    | md5                              | last_modified                 |
|------------------------------------------------------+---------+----------------------------------+-------------------------------|
| snfk_internal_stage/JC-202105-citibike-tripdata.csv.gz | 1214736 | d36282f3e8eaae8880bdf525801dc664 | Mon, 12 Jul 2021 19:44:23 GMT |
+------------------------------------------------------+---------+----------------------------------+-------------------------------+
1 Row(s) produced. Time Elapsed: 0.276s
DBADMIN#COMPUTE_WH@ANALYTICS_DB.CITIBIKE>
```

Figure 12-32. *List Command*

This shows that the file has been successfully copied from the client machine with the PUT command in SnowSQL.

Load Data with the COPY Command

Run a COPY command with appropriate parameters. You notice that we are not using a File Format object here. Instead, parameters are directly used with the COPY command:

e.g. > COPY INTO CITIBIKE_NYC FROM @SNFK_INTERNAL_STAGE FILE_FORMAT=CSV_FILE_FORMAT;

Figure 12-33 shows a COPY command to load data from files in the Internal Stage into a table.

```
DBADMIN#COMPUTE_WH@ANALYTICS_DB.CITIBIKE>
DBADMIN#COMPUTE_WH@ANALYTICS_DB.CITIBIKE>copy into CITIBIKE_NYC
                                from @snfk_internal_stage
                                file_format = (type= CSV  field_delimiter=    skip_header=1
                                empty_field_as_null=true
                                record_delimiter= '\n');
+-----------------------------------------------------+--------+-------------+-------------+-------------+-------------+-------------+------------------+-----------------------+------------------------+
| file                                                | status | rows_parsed | rows_loaded | error_limit | errors_seen | first_error | first_error_line | first_error_character | first_error_column_name |
|-----------------------------------------------------+--------+-------------+-------------+-------------+-------------+-------------+------------------+-----------------------+------------------------|
| snfk_internal_stage/JC-202105-citibike-tripdata.csv.gz | LOADED | 45043       | 45043       | 1           | 0           | NULL        | NULL             | NULL                  | NULL                   |
+-----------------------------------------------------+--------+-------------+-------------+-------------+-------------+-------------+------------------+-----------------------+------------------------+
1 Row(s) produced. Time Elapsed: 1.824s
DBADMIN#COMPUTE_WH@ANALYTICS_DB.CITIBIKE>
```

Figure 12-33. *Copy Command*

Verify the Results

Once the COPY command runs successfully, verify the results.

Figure 12-34 shows results with number of rows from the table.

```
DBADMIN#COMPUTE_WH@ANALYTICS_DB.CITIBIKE>
DBADMIN#COMPUTE_WH@ANALYTICS_DB.CITIBIKE>SELECT COUNT(*) FROM CITIBIKE_NYC;
+-----------+
| COUNT(*)  |
|-----------|
|     45043 |
+-----------+
1 Row(s) produced. Time Elapsed: 0.543s
DBADMIN#COMPUTE_WH@ANALYTICS_DB.CITIBIKE>
```

Figure 12-34. *Query and Result*

Summary

Snowflake is a cloud data platform that supports various workloads with its modern architecture and rich set of features. Snowflake is built for the cloud from the ground up, and it supports all major cloud providers (AWS, GCP, Azure).

There are different ways to load data into Snowflake with its native tools/commands or with third-party tools. The web UI provides an easy way to create objects and load smaller file(s) through a wizard. SnowSQL CLI provides a secure and easy way to copy and load data from supported client machines to Snowflake tables.

These two options don't depend on cloud provider storage (External Stages) and can be used to securely transmit and load data into tables with parallelism and efficiency.

CHAPTER 13

Unloading Data

Please make sure you review Chapter 12 before you proceed further. You should be familiar with Snowflake stages, navigating through the Snowflake web interface, and the SnowSQL command-line interface (CLI) tool. You also need to download and install the SnowSQL CLI tool onto your local machine. In this chapter, we will discuss unloading data from Snowflake tables and results from SQL statements into files with the COPY INTO <location> command and the SnowSQL CLI GET command.

The process of unloading data is very similar to loading data except it's reversed. We will keep this simple and make use of datasets that are provided by Snowflake. Let's dig in!

Sample Dataset

With every Snowflake account, there are sample datasets for TPC-H and TPC-DS in the SNOWFLAKE_SAMPLE_DATA database. You don't need to be familiar with the datasets, entities, and relationships; but if you are curious and would like to explore and understand in detail, you can visit the following URLs and learn more about them:

TPC-H: www.tpc.org/tpch/
TPC-DS: www.tpc.org/tpc_documents_current_versions/pdf/tpc-ds_v2.5.0.pdf

In our case, we are going to use the TPC-H dataset. Figure 13-1 shows the entity relationship diagram for the TPC-H dataset.

© Frank Bell, Raj Chirumamilla, Bhaskar B. Joshi, Bjorn Lindstrom, Ruchi Soni, Sameer Videkar 2022
F. Bell et al., *Snowflake Essentials*, https://doi.org/10.1007/978-1-4842-7316-6_13

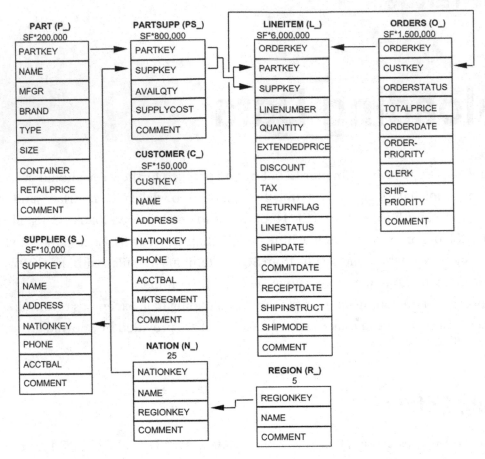

Legend:

- The parentheses following each table name contain the prefix of the column names for that table;
- The arrows point in the direction of the one-to-many relationships between tables;
- The number/formula below each table name represents the cardinality (number of rows) of the table. Some are factored by SF, the Scale Factor, to obtain the chosen database size. The cardinality for the LINEITEM table is approximate (see Clause 4.2.5).

Figure 13-1. *Entity Relationship Diagram*

Snowflake provides four different schemas with various sizes in terms of rows for customers to test scaling factors. We will focus on the TPCH_SF10 schema to read from a table/query and unload data into files with the COPY INTO <location> command onto the Internal Stage and download the files onto the local machine with the GET command.

Setting Up the Context

If you already have a Snowflake user account, then you can follow the steps to unload data into file(s). All system-defined roles have access to sample datasets in the SNOWFLAKE_SAMPLE_DATA database.

We are not going to create any objects like databases, schemas, or tables. We just need to create a Named Internal *Stage* object, where the unloaded files are staged. We will set the DEMO_DB database, PUBLIC schema, and SYSADMIN role as our worksheet and CLI context.

Log In to Your Snowflake Account

Log into your Snowflake account with your username and password. If you don't already have an account, please review instructions on how to sign up for a trial account in the previous chapter. Figure 13-2 shows the sign-in page for a Snowflake account.

Figure 13-2. *Snowflake Sign-In page*

Create a Stage

Let us create a stage object in the PUBLIC schema within the DEMO_DB database. Click the Databases icon 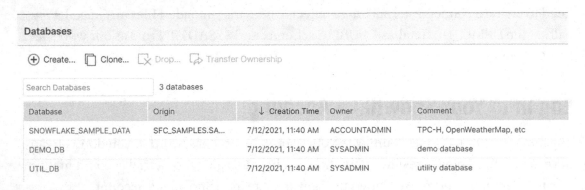 near the top-left corner.

Figure 13-3 shows the Databases page with DEMO_DB listed.

Databases

⊕ Create...   ◻ Clone...   ⬜ Drop...   ⬜ Transfer Ownership

Search Databases	3 databases			
Database	Origin	↓ Creation Time	Owner	Comment
SNOWFLAKE_SAMPLE_DATA	SFC_SAMPLES.SA...	7/12/2021, 11:40 AM	ACCOUNTADMIN	TPC-H, OpenWeatherMap, etc
DEMO_DB		7/12/2021, 11:40 AM	SYSADMIN	demo database
UTIL_DB		7/12/2021, 11:40 AM	SYSADMIN	utility database

Figure 13-3. *Databases Page with a List of Existing Databases*

From the list of databases, click the DEMO_DB database name and then select the Stages tab. Figure 13-4 shows the list of stages within DEMO_DB (at this point it's empty).

Databases > **DEMO_DB**

Tables   Views   Schemas   **Stages**   File Formats   Sequences   Pipes

⊕ Create...   ◻ Clone...   ✎ Edit...   ⬜ Drop...   ⬜ Transfer Ownership

Stage	Schema	Location	Creation Time ▼	Owner	Comment

Figure 13-4. *Stages Within the DEMO_DB Database*

To create a new stage, click the ⊕ Create... icon; and in the pop-up, select the ***Snowflake Managed*** location option and click ⬜ Next . Figure 13-5 shows location options (internal and external) for files to be staged.

Figure 13-5. *Stage Options*

In the next screen, enter values for the parameters. Name should be OUTBOUND_
INT_STAGE (required). For Schema Name the default PUBLIC value should have been
selected in the dropdown (if not, pick the value manually from the dropdown).
Comment should be Outbound Internal Named Stage(optional). Click Finish .
Figure 13-6 shows the *Create Stage* input values.

Create Stage

Staged files will be stored in a Snowflake managed stage

Name *	OUTBOUND_INT_STAGE
Schema Name	PUBLIC
Comment	Outbound Internal Named Stage

Show SQL   Cancel   Back   Finish

Figure 13-6. *Stage Input Values*

Once the stage is created, you should see the stage details. Figure 13-7 shows the
stage details that were just created.

Databases › **DEMO_DB**

Tables	Views	Schemas	**Stages**	File Formats	Sequences	Pipes

⊕ Create...   ☐ Clone...   ☑ Edit...   ☒ Drop...   ⤶ Transfer Ownership

Stage	Schema	Location	Creation Time ▼	Owner	Comment
OUTBOUND_INT_STAGE	PUBLIC	Snowflake	6:48:04 PM	SYSADMIN	Outbound Internal Named Stage

Figure 13-7. *Stage (OUTBOUND_INT_STAGE) Details*

The **_Location_** column value _Snowflake_ indicates that this stage is a Snowflake-managed Internal Stage.

Instead of creating the stage through the wizard, you can run the following DDL in a worksheet. It's an easy alternative approach, and you will get the same result:

```
CREATE STAGE "DEMO_DB"."PUBLIC".OUTBOUND_INT_STAGE COMMENT = 'Outbound
Internal Named Stage';
```

COPY INTO <location> and GET Commands

COPY INTO <location> Command

COPY INTO <location> is a SQL command, which is used to (bulk) unload data from tables into file(s) with parallelism (i.e., when extracted to multiple files).

With the COPY INTO <location> command, data can be unloaded into files from tables or from SQL statements (results) into Snowflake-managed stages (User, Table, or Internal Named), external locations, or Named External Stages that reference external locations (Amazon S3, Google Cloud Storage, or Microsoft Azure).

When files are staged into a Snowflake Internal Stage, then the GET command can be used to download the files onto other workstations or servers.

At the time of this writing, COPY INTO <location> supports unloading data into delimited text files of type CSV, TSV, etc. and of JSON format or PARQUET format. The SELECT statement must be used to unload data from a table. Snowflake by default compresses the files using gzip compression. By using the COMPRESSION parameter, you can specify any valid compression option. Currently, the options available for the COMPRESSION parameter are AUTO, GZIP, BZ2, BROTLI, ZSTD, DEFLATE, RAW_DEFLATE, and NONE. For the latest supported compression options, refer to the latest Snowflake documentation: `https://docs.snowflake.com/en/sql-reference/sql/copy-into-location.html`.

Splitting into Multiple Files

The COPY INTO <location> SINGLE parameter helps to specify whether the COPY command should unload data into a single file or multiple files. The default value for SINGLE is false.

Number (> 0) specifies the upper size limit (in bytes) of each file to be generated in parallel per thread. Note that the actual file size and number of files unloaded are determined by the total amount of data and number of nodes available for parallel processing.

The MAX_FILE_SIZE parameter specifies the upper size limit in bytes of each file generated in parallel per thread. The default setting for MAX_FILE_SIZE is 16777216 bytes (16 MB). The value for this parameter can be set up to 5 GB, based on the cloud provider's supported size limits. The number of files unloaded and the actual file size of them are determined by the number of node write threads available for the operation, the total amount of dat, and MAX_FILE_SIZE settings. All the files that are created during the operation may not have the same size.

GET Command

The Snowflake GET command supports downloading files from Snowflake's Internal Stages, that is, Table, User, and Named Stages, to local file systems on a client machine. Snowflake does not support downloading files from External Stages with the GET command. The GET command is not supported to run explicitly in the Snowflake web interface; but it is supported with the SnowSQL CLI, ODBC and JDBC drivers, and Python Connector. It always encrypts and compresses files before downloading.

Unloading Data

In this section we explore unloading data from a table as well as SELECT statements (results) into the Internal Stage we just created. To start with, Click ⧁ Worksheets from the top ribbon menu. In the worksheet, set the context. Figure 13-8 shows the context menu and the values.

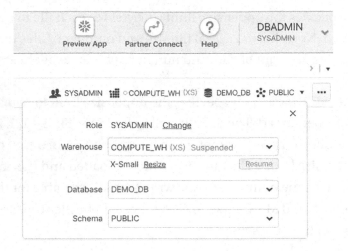

Figure 13-8. *Worksheet Context*

From a Table

First, let's list if there are any files in Internal Stage OUTBOUND_INT_STAGE. Figure 13-9 shows the Snowflake LIST command and the results.

Figure 13-9. *List of Files from Internal Named Stage*

Now let's run a COPY INTO <location> command to unload data from the CUSTOMER table in the TPCH_SF10 schema in the SNOWFLAKE_SAMPLE_DATA database. Figure 13-10 shows the COPY INTO <location> command and the results.

Figure 13-10. *Unload Command from a Table*

Note that this COPY INTO <location> command unloads data into the customer folder within the outbound_int_stage Internal Stage in our example. If you do not specify the SINGLE or MAX_FILE_SIZE parameter with the command, then it will use the default values as described previously. Since the default value for the COPY INTO <Location> SINGLE option is false, we can expect multiple files. Let's see the file attributes by using the LIST command.

From a SQL Statement

Now let's use a query to unload results into the *mincostsupplier* folder in the *outbound_int_stage* Internal Stage. Use the following TPC-H query, which outputs minimum-cost suppliers for a particular part in a specific region. Let's not worry about the query and the logic. Our intention here is to just unload the results:

```
-- Minimum Cost Supplier Query
copy into @outbound_int_stage/mincostsupplier.csv
from
(select
      s_acctbal,
      s_name,
      n_name,
      p_partkey,
```

```
        p_mfgr,
        s_address,
        s_phone,
        s_comment
from
        snowflake_sample_data.tpch_sf10.part,
        snowflake_sample_data.tpch_sf10.supplier,
        snowflake_sample_data.tpch_sf10.partsupp,
        snowflake_sample_data.tpch_sf10.nation,
        snowflake_sample_data.tpch_sf10.region
where
        p_partkey = ps_partkey
        and s_suppkey = ps_suppkey
        and p_size = 10
        and p_type like 'ECONOMY BURNISHED COPPER'
        and s_nationkey = n_nationkey
        and n_regionkey = r_regionkey
        and r_name = 'AMERICA'
        and ps_supplycost = (
            select
                min(ps_supplycost)
            from
                snowflake_sample_data.tpch_sf10.partsupp,
                snowflake_sample_data.tpch_sf10.supplier,
                snowflake_sample_data.tpch_sf10.nation,
                snowflake_sample_data.tpch_sf10.region
            where
                p_partkey = ps_partkey
                and s_suppkey = ps_suppkey
                and s_nationkey = n_nationkey
                and n_regionkey = r_regionkey
                and r_name = 'AMERICA'
        )
order by
        s_acctbal desc,
        n_name,
```

```
    s_name,
    p_partkey
)
single=TRUE;
```

Note that in this code example, now the COPY INTO <Location> parameter SINGLE is set to TRUE. In this case we can expect one single output file. Also, the output file name is set in the copy command. Figure 13-11 shows the unload command with a SQL statement.

Figure 13-11. *Unload Command and the Results*

Verify the Results

Now, let's list the files we unloaded for use cases. Figure 13-12 shows the LIST command and the results after executing the COPY INTO <location> command to unload the data from a table.

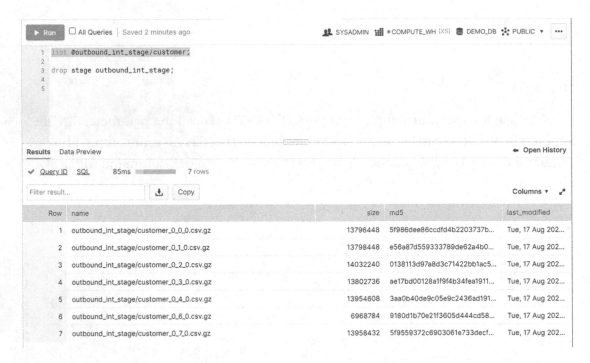

Figure 13-12. *List of Files Unloaded to outbound_int_stage/customer*

Figure 13-13 shows the LIST command and the results after executing the COPY INTO <location> command from a SQL statement.

Figure 13-13. *List of Files Unloaded to outbound_int_stage/mincostsupplier*

We have the unloaded data into the outbound_int_stage Internal Stage from a table and a SQL statement into *customer* and *mincostsupplier* folders, respectively.

Download Data via SnowSQL

Now let's use SnowSQL to download files locally onto the workstation using the GET command.

Log In and Set Context

To log in to snowflake through SnowSQL, open a command prompt and run the `snowsql` cli command.

We are going to use -a (account name) and -u (username) parameters with SnowSQL. When prompted, enter the password:

e.g. > snowsql -a yxxxxxx.us-east-1 -u XXXXX

The account name could be different based on your specific account cloud provider and region. The account name is simply the Snowflake account URL without the snowflakecomputing.com domain. Once you log in to SnowSQL, set the context with user role, database, and schema. Figure 13-14 shows snowsql command-line login information and context setting with use commands.

```
C:\Users>snowsql -a y          .us-east-1 -u DBADMIN
Password:
* SnowSQL * v1.2.17
Type SQL statements or !help
DBADMIN#COMPUTE_WH@(no database).(no schema)>use role SYSADMIN;
+----------------------------------+
| status                           |
|----------------------------------|
| Statement executed successfully. |
+----------------------------------+
1 Row(s) produced. Time Elapsed: 0.134s
DBADMIN#COMPUTE_WH@(no database).(no schema)>use database DEMO_DB;
+----------------------------------+
| status                           |
|----------------------------------|
| Statement executed successfully. |
+----------------------------------+
1 Row(s) produced. Time Elapsed: 0.122s
DBADMIN#COMPUTE_WH@DEMO_DB.PUBLIC>use schema PUBLIC;
+----------------------------------+
| status                           |
|----------------------------------|
| Statement executed successfully. |
+----------------------------------+
1 Row(s) produced. Time Elapsed: 0.127s
DBADMIN#COMPUTE_WH@DEMO_DB.PUBLIC>
```

Figure 13-14. *Login with snowsql cli and Context Setting*

List Files from Internal Stage

Similar to the web user interface, you can easily list files from the Snowflake Internal Stage with the LIST command in SnowSQL. Figure 13-15 shows an example of the command to list Named Internal Stage files in SnowSQL CLI.

```
DBADMIN#COMPUTE_WH@DEMO_DB.PUBLIC>list @outbound_int_stage;
+---------------------------------------------+----------+----------------------------------+---------------------------------+
| name                                        |     size | md5                              | last_modified                   |
|---------------------------------------------+----------+----------------------------------+---------------------------------|
| outbound_int_stage/customer_0_0_0.csv.gz    | 13796448 | 5f986dee86ccdfd4b2203737bda30ce8 | Tue, 17 Aug 2021 02:14:49 GMT   |
| outbound_int_stage/customer_0_1_0.csv.gz    | 13798448 | e56a87d559333789de62a4b054c6ba05 | Tue, 17 Aug 2021 02:14:49 GMT   |
| outbound_int_stage/customer_0_2_0.csv.gz    | 14032240 | 0138113d97a8d3c71422bb1ac5a54faa | Tue, 17 Aug 2021 02:14:49 GMT   |
| outbound_int_stage/customer_0_3_0.csv.gz    | 13802736 | ae17bd00128a1f9f4b34fea19113979b | Tue, 17 Aug 2021 02:14:49 GMT   |
| outbound_int_stage/customer_0_4_0.csv.gz    | 13954608 | 3aa0b40de9c05e9c2436ad191ee35720 | Tue, 17 Aug 2021 02:14:49 GMT   |
| outbound_int_stage/customer_0_6_0.csv.gz    |  6968784 | 9180d1b70e21f3605d444cd585bc5689 | Tue, 17 Aug 2021 02:14:49 GMT   |
| outbound_int_stage/customer_0_7_0.csv.gz    | 13958432 | 5f9559372c6903061e733decf8c738bf | Tue, 17 Aug 2021 02:14:49 GMT   |
| outbound_int_stage/mincostsupplier.csv      |    11488 | c41dd1c0f22c35688043d2066220e52f | Tue, 17 Aug 2021 02:17:20 GMT   |
+---------------------------------------------+----------+----------------------------------+---------------------------------+
8 Row(s) produced. Time Elapsed: 0.557s
DBADMIN#COMPUTE_WH@DEMO_DB.PUBLIC>
```

Figure 13-15. *File Listing from Named Internal Stage*

Download Files with the GET Command

Run a GET command to copy files from the Internal Stage to your local directory:

```
GET @outbound_int_stage file://c:\temp;
```

The GET command will encrypt and copy the files from your Snowflake Internal Stage to a local Temp directory. Figure 13-16 shows the execution of the GET command to copy files to the Temp directory on a client machine from the Named Internal Stage called outbound_int_stage.

```
DBADMIN#COMPUTE_WH@DEMO_DB.PUBLIC>GET @outbound_int_stage file://c:\temp;
mincostsupplier.csv(0.01MB): [##########] 100.00% Done (0.146s, 0.08MB/s).
customer_0_7_0.csv.gz(13.31MB): [##########] 100.00% Done (1.888s, 7.05MB/s).
customer_0_4_0.csv.gz(13.31MB): [##########] 100.00% Done (5.682s, 2.34MB/s).
customer_0_0_0.csv.gz(13.16MB): [##########] 100.00% Done (6.284s, 2.09MB/s).
customer_0_6_0.csv.gz(6.65MB): [##########] 100.00% Done (8.076s, 0.82MB/s).
customer_0_1_0.csv.gz(13.16MB): [##########] 100.00% Done (10.012s, 1.31MB/s).
customer_0_2_0.csv.gz(13.38MB): [##########] 100.00% Done (10.857s, 1.23MB/s).
customer_0_3_0.csv.gz(13.16MB): [##########] 100.00% Done (10.901s, 1.21MB/s).
+----------------------+----------+------------+----------+
| file                 |     size | status     | message  |
|----------------------+----------+------------+----------|
| customer_0_0_0.csv.gz | 13796447 | DOWNLOADED |          |
| customer_0_1_0.csv.gz | 13798438 | DOWNLOADED |          |
| customer_0_2_0.csv.gz | 14032235 | DOWNLOADED |          |
| customer_0_3_0.csv.gz | 13802724 | DOWNLOADED |          |
| customer_0_4_0.csv.gz | 13954595 | DOWNLOADED |          |
| customer_0_6_0.csv.gz |  6968768 | DOWNLOADED |          |
| customer_0_7_0.csv.gz | 13958428 | DOWNLOADED |          |
| mincostsupplier.csv  |    11480 | DOWNLOADED |          |
+----------------------+----------+------------+----------+
8 Row(s) produced. Time Elapsed: 13.085s
DBADMIN#COMPUTE_WH@DEMO_DB.PUBLIC>
```

Figure 13-16. *GET Command to Download Files from Internal Stage*

Let's view the list of files downloaded from Snowflake's Internal Stage to the local Temp directory. Figure 13-17 shows the list of files downloaded with the GET command.

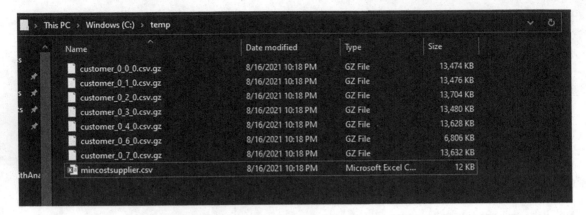

Figure 13-17. *List of Files from Internal Stage*

We now successfully unloaded data from a table and SQL statement into the Internal Stage and downloaded the files from the Internal Stage to a workstation with the GET command.

Summary

Snowflake makes unloading data to Internal and External Stages easy and simple. It is always recommended to use data sharing capabilities that are unique to the Snowflake Data Cloud first to share data with simple and controlled operations. Sometimes you need to export or unload data though for your business use cases. As we reviewed, it is easy to bulk unload data from Snowflake tables and SQL statements into files and download data. SnowSQL CLI provides a secure and easy way to download data into supported client machines. The options reviewed in this chapter don't depend on cloud provider storage (External Stages) and can be used to securely transmit and unload data from tables or SQL statements with parallelism and efficiency and download data to supported clients.

Data Sharing, Data Exchanges, and the Snowflake Data Marketplace

In this chapter, we will cover Snowflake's data sharing features. Data sharing is one of Snowflake's major differentiators from all other RDBMSs and cloud data systems. Data sharing is really what takes Snowflake from just another cloud database, data warehouse, or data lake to the larger concept of the Snowflake Data Cloud. Snowflake's unique metadata-driven architecture based on separate compute and storage allows it to become a full data cloud solution. The Snowflake Data Cloud can be used as a single source of truth for an organization, multiple organizations, or even the entire world of data users with both open and private sets of data. If this data sharing functionality continues to gain traction, it is a very powerful concept to think that one single dataset could be shared by thousands or millions of organizations or people including both public and private companies and governments without having to make copies of the data.

In the past, humans and their corporations have duplicated most of their common data due to both on-premise and cloud data scaling challenges as well as privacy and data ownership concerns. The reality is that human interconnectivity and a push for more automation and simplification will push companies toward sharing at least some datasets. I mean, why should we have millions and millions of different data tables for common datasets such as lists of countries, states, and cities throughout the world? The current data copying techniques result in additional unnecessary duplication of storage and increase energy consumption that impacts climate change. At the same time, once

© Frank Bell, Raj Chirumamilla, Bhaskar B. Joshi, Bjorn Lindstrom, Ruchi Soni, Sameer Videkar 2022
F. Bell et al., *Snowflake Essentials*, https://doi.org/10.1007/978-1-4842-7316-6_14

datasets are shred and used by many organizations, those organizations all become dependent on the quality and security of that dataset. Within this chapter, we will cover the three conceptual data sharing mechanisms of regular data sharing, Private Data Exchanges, and the Public Snowflake Data Marketplace.

The Snowflake Private Data Exchanges and Public Data Marketplace are just the beginning of the functionality to share open and private datasets securely and in near real time. Currently the marketplace is growing quickly but is still relatively immature, but it has great potential. I have worked with Snowflake Corporation from the beginning of their Data Exchanges in 2018 to provide guidance for the improvement of data sharing on both Private Data Exchanges and the Data Marketplace. The Snowflake Data Marketplace has really improved and grown since its debut in June 2019.

Data Marketplaces overall though are very new concepts, and for many data professionals it still is an extremely large paradigm shift to go from localized copies of data to globally shared data sources. Data and business professionals would need to move from having their corporate silos of data and their corporate controlled dimensions of data to a more interconnected collaborative shared data architectural concept. It will take time for data professionals, business professionals, and corporations to change. Let's start with reviewing what data sharing is and how it works on Snowflake.

Data Sharing History
What Is Data Sharing?

Let's make sure we understand what data sharing is overall before we get into how it has been revolutionized on the Snowflake Data Cloud. The concept of data sharing really has been around since before there was even electronic data to share. Data is information that has been made into semi-structured or delimited form, which is efficient for processing and decision making.

We will focus on electronic data sharing. Since the beginning of computers, data processing created results or data outcomes that could be shared. Initially, this was done by printouts or raw media. It then evolved into the copying of structured or semi-structured data through files or application programming interfaces (APIs). These mechanisms are still where the bulk of data sharing is done today. Compared to Snowflake's data sharing though, they are less efficient and have more latency due to their overall reliance on the copying of data.

Data Copies Everywhere

Data copies, copies, copies, everywhere. Hey, I get it. We humans want control. We wanted to own our data and know that no one can take it from us. The problem is that copied data without any control and organization is an organizational nightmare, which can potentially lead to truly bad outcomes.

Before Snowflake's groundbreaking data sharing feature came to the market, we data professionals for tens of years built millions of data copying jobs across the world. We copied data within one database. We aggregated data to scale querying better. We copied data from one database vendor to another one to try and achieve better performance or just easier usability or security across business units and/or partners. We copied data into CSV files and opened them in text editors or Excel. We copied data into tab-separated files and other delimited files everywhere. The challenge with all this copying (and this is still very predominant with data right now) is, how do you know what the latest dataset with the correct data is? When you have multiple copies of data (often which is ungoverned), this becomes much more challenging. It is all too easy for any data analyst or decision maker to be analyzing data that is inaccurate and out-of-date.

The BI Tool Extract Effect and Data Governance and Integrity Impacts

To further compound this data scaling and copying problem, many of the BI vendors built out extraction functionality and capabilities to differentiate themselves and speed up querying on the databases they retrieved data from. Most of these enhancements were complex layers creating automated copied aggregates of data to speed up queries. Tableau, Domo, Qlik, and others created proprietary ways to increase performance by mainly extracting datasets or placing proprietary indexes and speed enhancement mechanisms as well. One of the really challenging problems with all these implementations and usages of these tools is that they were designed at a time before Snowflake and other scalable cloud data systems and now with today's cloud speed and scale these old implementations just create complexity and outdated data. In large organizations though, this is a recipe for Data-Driven Disasters (DDD). Too often, two analysts or C-level executives can have two different extracts of what should be the same dataset. This leads to business confusion and DDD.

How much better will a business run if it knows all its automations and analysis systems use data that is governed, secured, validated, consistent, and up-to-date? If you think about it, data and the decisions derived from it are what have won and lost battles and wars and significantly changed our world. Knowing exactly where your enemy was and what resources they had provided you or your opponent extreme advantages. In business, knowing that your data is accurate and up-to-date is incredibly important as well. Any decisions based on inaccurate data can lead to seriously bad consequences.

Snowflake has become the pioneer for this new form of secure, instantaneous, non-copied dependent form of data sharing that is revolutionizing organizations throughout the world.

Data Sharing on the Snowflake Data Cloud
Snowflake Data Sharing History

Snowflake had built the data sharing concept into their cloud data warehouse from the beginning but really started promoting it initially as a Data Sharehouse in June 2017. Snowflake at that time was relatively small and almost entirely focused on being the best cloud data warehouse, so while the technology was amazingly good, it had limited awareness in the overall database marketplace. When I discovered Snowflake at the beginning of 2018, I realized that the data sharing functionality was extremely powerful and differentiated from anything I had been doing in the last 25 years of working with RDBMS, NoSQL, and other data systems. It was the first time my teams didn't have to create complex data pipelines based on the copying of data in delimited or semi-structured files or using APIs to transfer data within a company or outside of a company. Let's go through why Snowflake's data sharing is so revolutionary in the following.

The Business Value and Power of Instantaneous Data Sharing

Figure 14-1 shows how Snowflake's unique concept of sharing data without copying has made it so much easier to securely and instantly share governed data. Snowflake data sharing transformed business to be able to share data without the overhead and complexity of copying the data into files and sharing via FTP, SFTP, email, or API. The age of copying data and exports can be left behind for an overall better data collaboration experience.

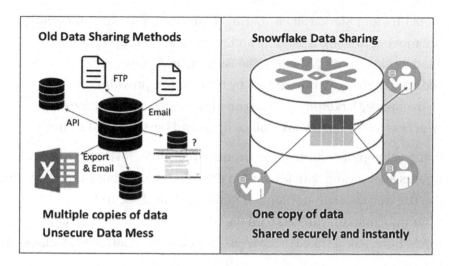

Figure 14-1. *The Power of Instantaneous Data Sharing*

By enabling such a low-cost and low friction way to share data, Snowflake has empowered organizations to be able to share faster, in larger volumes, and at a much lower cost. Some of the major business/organizational challenges Snowflake's low-friction data sharing solves today are the following use cases:

1. Cross-Enterprise Data Sharing: In many of the Fortune 100 we have consulted with, there still unfortunately remain many silos of data. The easiest way to break down those silos with scale, speed, and no copying has been to move to the Snowflake Data Cloud and enable near-real-time data sharing with governance.

2. Partner/Extranet Data Sharing: (This functionality can be enabled with basic data sharing we describe in this section or with more administrative control and ease of use with the Private Data Exchange functionality we cover later). Most organizations have many suppliers and other constituents that can benefit from near-real-time securely shared governed data with them. Costly data extranets, EDI (Electronic Data Interchange), APIs, and data pipelines can be removed or simplified and augmented with Snowflake's data sharing functionality.

3. Data Provider Marketplace Sharing: Many new and old companies are moving to share their data and monetize it through Data Marketplace's data providers like FactSet, and others have reduced their cost to share data by becoming a provider on Snowflake's Marketplace. Data consumers have enjoyed the ease of access to the Data Marketplace open and personalized shares.

Overall, the value proposition of secure, near-real-time, instantaneous, low-cost data sharing was an amazing leap forward in data-driven collaboration. It continues to revolutionize the way data can be collaborated and shared securely.

For further information, you can read the original "The Power of Instantaneous Data Sharing" article here: `www.linkedin.com/pulse/power-instantaneous-data-sharing-frank-bell/`.

Data Sharing Market Adoption

When I published the article "The Power of Instantaneous Data Sharing" back in 2018, I naively thought that the data sharing business case was so obvious and would move business and technology users to Snowflake's data sharing much more rapidly than I'd seen. I could see how this feature completely removed massive friction from the data sharing between corporations internally and their partners and through marketplaces. The reality is for many organizations, change is a major mental shift both in learning and trusting new technical methods. Education takes time especially for more conservative industries and organizations. In 2021, the cloud seems obvious now; it has been around for over 10+ years. In 2010, most organizations would not move to the cloud at all. I assume Snowflake's data sharing may follow a similar pattern except given the network effects of data sharing, it should have a faster adoption unless additional competitors are able to duplicate the feature.

How Data Sharing Works on Snowflake

Now that we know the pain of prior ways to share data by copying, let's dive into how Snowflake's unique metadata-driven separated storage architecture enables data sharing securely in near real time. Figure 14-2 provides an example of data sharing happening in both Region 1 and Region 2. Snowflake's unique immutable partitioned data architecture, which separates storage and compute, allows to share the data in near real time without any copying of data [unless replication is required to an account in another region]. The data provider can easily create a share that can share either table, view, or user function objects from a database within the data provider's account. The data provider then grants access to that share object from their account. Figure 14-2 shows how a data provider with Data Provider Account 1 creates a share and grants access to ONLY Consumer Account A and Consumer Account B in that example. If the data provider needs to share data in another region, then they will replicate the data from one Snowflake region to another as shown in Figure 14-2 with the replication arrow from Data Provider Account 1 to Data Provider Account 2. From their second account in a different region, the data provider in this example has granted access to the data share to Consumer Accounts X and Y.

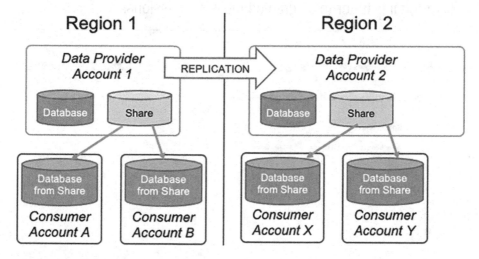

Figure 14-2. *Snowflake Secure Data Sharing*

How to Create Data Shares and Access Data Shares on Snowflake

Now that we know how the Snowflake data sharing technology works and the inefficiencies of other data sharing techniques, let's go through how you can easily set up standard data sharing on Snowflake. It is relatively easy to set up Snowflake data shares either through Snowflake SHARE definition commands or through their web interface.

How to List Snowflake Data Shares – Shared Data

To see a listing of incoming data shares [or shared data] in your account, you go to this URL with your region and account name details. Remember, prior to replication, you can only create a data share to be shared within the same region:

`https://app.snowflake.com//[your-region-name]/[your-account-name]/shared/shared-with-me`

Tip   If you see Figure 14-3 with the message of *Permissions Required*, that means you do not have enough permissions to list or create data shares. Contact your Account Administrator to be granted these permissions.

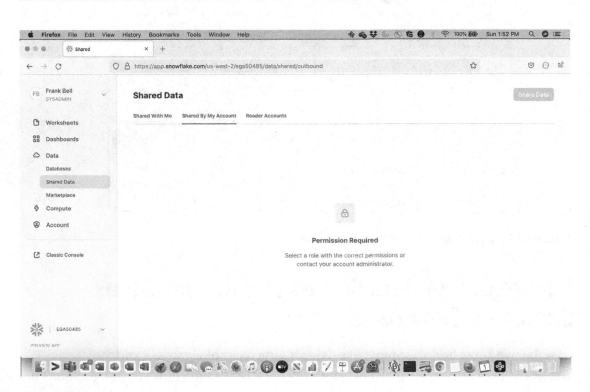

Figure 14-3. *Data Share Listings – No Access (Snowsight)*

Figure 14-4 shows the Snowsight (Preview App) for the *Shared With Me* view when you have permissions for data share listings.

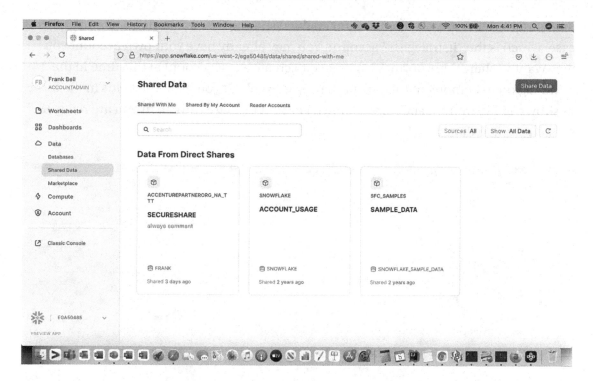

Figure 14-4. *Data Share Listings – Classic Console*

If you are still using the Snowflake Classic Console, we previously showed you how to view listings in Figure 4-12 on the Classic Console.

How to Create a Snowflake Data Share

Setting Up a Data Share on the Snowflake Web Interface [Snowsight Example]

Now that you are in the *Shared Data* listings area, all you need to do to set up a data share for another business unit or partner with another Snowflake account or your own Reader Accounts is click the blue *Share Data* button in the upper right of the screen. This brings up the form in Figure 14-5, which displays a filled-in Create Share form based on our CITIBIKE dataset. Assuming you already created the CITIBIKE database, you can easily navigate to that and to the TRIPS table within it to replicate the settings in Figure 14-5. Remember, when you create a share from Snowsight, you must have a data object ready to be shared (table or secure view). Otherwise, the *Create Share* button will not be enabled, and you cannot even search for an account to share with. We also show in Figure 14-5 how we searched for and enabled sharing for our sharing account in this case, which is the "itstrategists" account. You may already have your own account you may want to share with, or you would need to create a test account at this time in the same region to demonstrate this data sharing example. If you fill in the fields in the form as shown in Figure 14-5 and have a known account to share with, then you are ready to create your first data share!

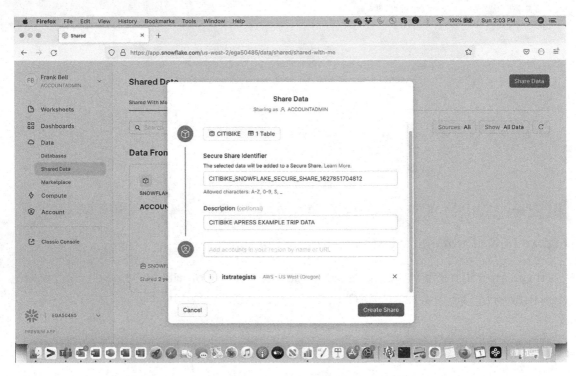

Figure 14-5. *Creating a Snowflake Data Share – Snowsight*

If you are still using the Snowflake Classic Console, we showed you how to create a new data share back in Figure 4-13. When you click *Create Share* in the screen shown previously, then your data share is activated for the account you shared with immediately.

Consumer Account: Getting Access to the New Data Share

Now that you have an active data share enabled by your data provider account, you need to go to the Snowflake data consumer account to which you just gave access to this CITIBIKE data share. When you log in to this data consumer account, go to Data ➤ Shared Data on the Snowsight screen. Figure 14-6 shows what the data consumer interface will look like if you have the proper role enabled that has access. Notice the *Ready to Get* data share has the CITIBIKE listing in a rectangle. Again, if you see Figure 14-3 on this new account, Data ➤ Shared Data screen with *Permissions Required*, then you typically must change roles to the ACCOUNTADMIN role or equivalent or work with an Account Administrator to get access that you need. Otherwise, you do not have enough access to perform the following steps and get to the data share just shared with this account. Assuming you can see a screen like Figure 14-6, then toward the bottom you just need to click *Get shared data*. Remember, this is an entirely different account you are using. The one other way to provide access is to also create Reader Accounts, which we will cover in the following.

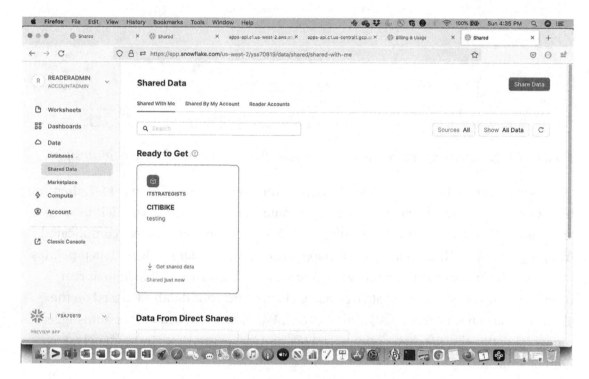

Figure 14-6. *Accessing a Snowflake Data Share – Snowsight*

Once you have clicked the *Get shared data* link, then Figure 14-7 will be displayed. This is where you set the new name of the actual database created from the data share you just received access to. Also, you can add permissions to which roles can access this data share. Remember that data shares are currently read only.

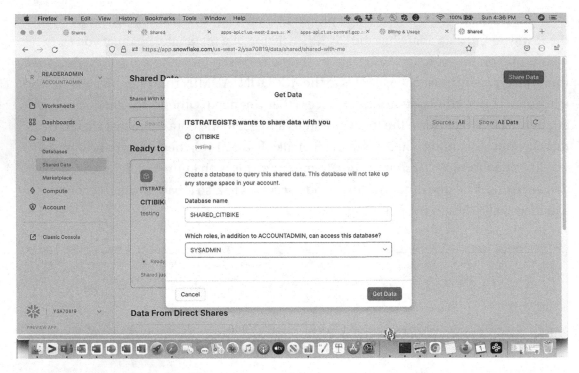

Figure 14-7. *Creating a Snowflake Database from the Data Share – Snowsight*

Once you click the blue *Get Data* button on the bottom right of Figure 14-7, then you have created your new database based on the data share provided to you. It is as simple and easy as that to create this read-only access from another account. No complicated API programming. No complex data file copying or complex data logic in data pipelines. Just a simple database and schema with tables and/or views accessible in near real time. You can easily go to the data tree navigation to find your database based on the share, or you can start writing SQL code in your Worksheets area and have autosuggest recognizing the new database, schema, tables, and/or secure views immediately. It is kind of amazing.

How to Create Reader Accounts

Let's say though that your partner or other business unit does not have a regular Snowflake account. Then Snowflake has created this awesome feature of Reader Accounts where you can easily create new Snowflake Reader Accounts to share data. Just remember, these accounts are paid for by the parent organization account that creates them. As mentioned previously, you must use best practices with security and warehouse design before giving access to these accounts. By default, there is no warehouse created in these accounts, but if you gave standard ACCOUNTADMIN or SYSADMIN rights, then the user who has those can easily create unbound warehouses, which are a huge data cost consumption risk. The best practice is to use the administrator account to create BOTH a reader warehouse and a reader resource monitor that will suspend based on a credit limit. Figure 14-8 displays the *New Reader Account* interface on Snowsight. Fill in the form with the new account name, comment, and administrator username and password and then click *Create Account*.

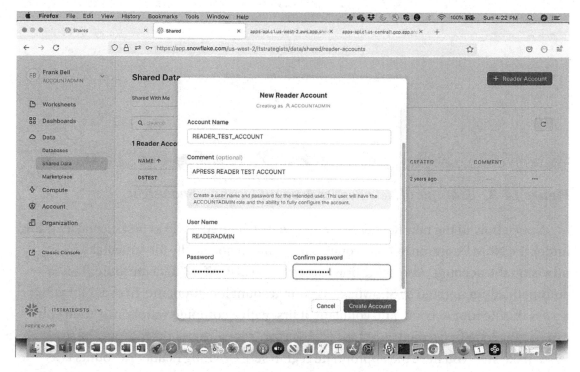

Figure 14-8. *Creating Reader Accounts – Snowsight*

If you are still using the Classic Console, then we showed how you can create Reader Accounts previously in Figure 4-9. Once you have clicked *Create Account* in a screen like Figure 14-8, then a new Reader Account listing screen (Figure 14-9) will appear.

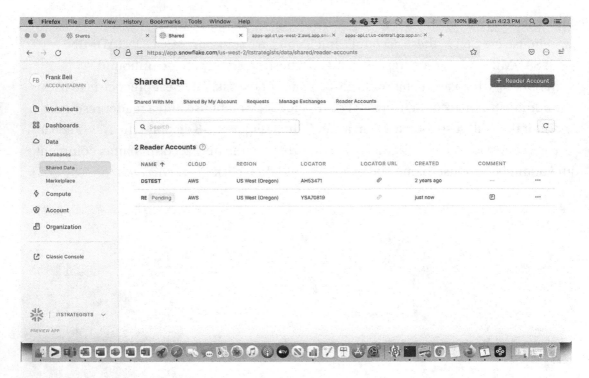

Figure 14-9. *Example of New Reader Account Listing – Snowsight*

Notice how in the new Reader Account you just created it will at first display Pending under the NAME column. It takes roughly a minute or so usually for Snowflake to provision the account. Also, write down or copy the LOCATOR data attribute. This is automatically generated by Snowflake as your account locator name. In order for you to access this Reader Account and configure it first with a compute warehouse, then you can use the locator account as `https://<locator-name>.snowflakecomputing.com` [also make sure you have the appropriate region set]. Creating Reader Accounts is as simple as that, but remember these are full Snowflake accounts and you must administer them and monitor them appropriately.

Setting Up a Data Share with Code

Now that you know how to set up a standard data share with the Snowsight web interface and grant access to it with either regular accounts or Reader Accounts or both, let's show how easy it is to do this in Snowflake definition language. Also, if you are doing shares at scale, this is much easier to scale via code. We will break out code based on if you are enabling the provider account or the consumer account.

Data Provider Account: Data Share Setup Code

```
/* Create the CITIBIKE Data Share and Grant appropriate access.  Remember,
for the share to work you must grant the correct access to the
shared objects
*/

CREATE SHARE "SS_CITIBIKE_DATA_SHARE" COMMENT='Citibike Data Share Demo';
GRANT USAGE ON DATABASE "ANALYTICS_DB" TO SHARE "SS_CITIBIKE_DATA_SHARE";
GRANT USAGE ON SCHEMA "ANALYTICS_DB"."CITIBIKE" TO SHARE "SS_CITIBIKE_
DATA_SHARE";
GRANT SELECT ON VIEW "ANALYTICS_DB"."CITIBIKE"."CITIBIKE_NYC" TO SHARE "SS_
CITIBIKE_DATA_SHARE";

/* Now that you have a share created, you need to GRANT access to the
Snowflake Account you want to share with.  The account must be in the
same region currently. Below is an example of granting access to my
ITSTRATEGISTS account.  You would replace ITSTRATEGISTS with your own
account.
*/

ALTER SHARE "SS_CITIBIKE_DATA_SHARE" ADD ACCOUNTS=ITSTRATEGISTS;
```

Data Consumer Account: Data Share Access Code

```
/* Go to the DATA CONSUMER Snowflake Account you just granted the share
access to and create a S_CITIBIKE Database based off of the share named
SS_CITIBIKE_DATA_SHARE. Remember, you need to be in the ACCOUNTADMIN
role at first to access the share and create a database from it. These
statements below will not work unless you are in the ACCOUNTADMIN role in
the Worksheet or equivalent
*/

CREATE DATABASE "S_CITIBIKE" FROM SHARE <ACCOUNT_NAME_OF_DATA_SHARE_
CREATOR>."SS_CITIBIKE_DATA_SHARE" COMMENT='Creating S_CITIBIKE database
from SS_CITIBIKE_DATA_SHARE';
```

/* Optional. Once you have created a database from the share then it is ONLY accessible typically to ACCOUNTADMIN role at first. In order to add other roles to access the read only Database built from the Share you GRANT access as done below. This grants all the privileges of the share to the role SYSADMIN   */

```
GRANT IMPORTED PRIVILEGES ON DATABASE "S_CITIBIKE" TO ROLE "SYSADMIN";
```

Data Provider Account: Creating Reader Accounts

/* CREATING A READER ACCOUNT which is a MANAGED Account that you the READER ACCOUNT CREATOR pay for usage.
*/

```
CREATE MANAGED ACCOUNT CITIBIKE_READER admin_name='READER_ADMIN', admin_
password='ENTER_SECURE_PWD', type=reader, COMMENT='Citibike Example Reader
Account';
```

Reader Account [Also, Data Consumer] Initial Setup: Creating Limits on Reader Compute (Warehouses)

/* This code below is my best practice suggestion for a newly created Reader Account in Snowflake. It is imperative that you go in as an administrator and create the warehouse with credit limits and that you also create users with limited query access. */

```
CREATE WAREHOUSE READER_WH WITH WAREHOUSE_SIZE = 'XSMALL' WAREHOUSE_TYPE
= 'STANDARD' AUTO_SUSPEND = 1 AUTO_RESUME = TRUE COMMENT = 'Snowflake
Solutions Data Sharing Warehouse';
```

/* Our best practice is always to create a Resource Monitor per warehouse. In this example below we created a READER_WH_RM resource monitor to monitor and suspend the READER_WH we just created above on the reader account. This resource monitor has a credit limit of 5 credits over one calendar month. The resource monitor named READER_WH_RM has three settings of 90% notify the ACCOUNTADMIN, at 95% of the 5 credits it will suspend after query at 90% and it will suspend immediately at 100% usage of 5 credits. Remember, resource monitors currently are very limited on who they can notify. Also, you must make sure you set yourself up for notifications on this Reader Account to get them. See Notifications setting at the reader account: https://<account_name>.snowflakecomputing.com/console#/preferences

```
*/
CREATE RESOURCE MONITOR "READER_WH_RM" WITH CREDIT_QUOTA = 5, frequency =
'MONTHLY', start_timestamp = 'IMMEDIATELY', end_timestamp = null
 TRIGGERS
 ON 95 PERCENT DO SUSPEND
 ON 100 PERCENT DO SUSPEND_IMMEDIATE
 ON 90 PERCENT DO NOTIFY;
ALTER WAREHOUSE "READER_WH" SET RESOURCE_MONITOR = "READER_WH_RM";
```

Reader Account [Also, Data Consumer] Initial Setup: Creating Users with Only Query Access to CITIBIKE Database

```
/* Now that you have created a warehouse that is constrained to 5 credits
let's create a reader account user which can only access the Citibike
Database based on the Data Share.
CREATE USER READER_QUERY_USER PASSWORD = 'ENTER_YOUR_PWD_HERE' COMMENT =
'Citibike Example Reader Query Only User' LOGIN_NAME = 'READER_QUERY_USER'
DISPLAY_NAME = 'READER_QUERY_USER' FIRST_NAME = 'FnameReader' LAST_NAME
= 'LnameReader' EMAIL = 'fname.lname@reader.com' DEFAULT_ROLE = "PUBLIC"
DEFAULT_WAREHOUSE = 'READER_WH' DEFAULT_NAMESPACE = 'ANALYTICS_DB' MUST_
CHANGE_PASSWORD = TRUE;

/* Notice.  We only GRANT this user the role of PUBLIC to make sure they
cannot create new warehouses or other functionality that read only partner
would not need.
*/
GRANT ROLE "PUBLIC" TO USER READER_QUERY_USER;

/* To make it easy for the user named READER_QUERY_USER with the PUBLIC
role to access the Database S_CITIBIKE then let's grant their PUBLIC
role access.
*/
```

```
GRANT IMPORTED PRIVILEGES ON DATABASE "S_CITIBIKE" TO ROLE "PUBLIC";

/*. Finally, you need to GRANT USAGE to READER_WH so the read only user can
run queries.  You do this by granting usage to the PUBLIC role assigned to
the READER_QUERY_USER.  Now this READER_QUERY_USER user can only use the
READER_WH with the 5 credits limit over the calendar month.  You can add
additional users to use the same warehouse as well by assigning them the
PUBLIC role.  Overall, your cost exposure now is 5 credits per month.
*/

GRANT USAGE ON WAREHOUSE "READER_WH" TO ROLE "PUBLIC";
```

Tip   It is incredibly important to configure your Reader Account before giving access to the reader users who will use it. You should either reuse the preceding code and alter it to meet your environment needs or use the web interface to execute some similar initial configuration. If you do not do this, then you are exposing your account to unconstrained Snowflake costs.

Data Exchanges

Once Snowflake realized the business value of basic data sharing, they proceeded to add functionality to allow for the concept of Private Data Exchanges. Snowflake Private Data Exchanges enable their owners to act similarly as Snowflake does with their Data Marketplace. The Snowflake Data Marketplace is just a Private Data Exchange where Snowflake is the actual Exchange Owner/Administrator. Snowflake's current Private Data Exchange offering has evolved for the Data Exchange Owner/Administrator to enable who can be a data provider and data consumer within this Private Data Exchange.

Creating Your Own Private Data Exchange Steps

Let's take you through the steps of setting up your own Private Data Exchange.

Prerequisites for Creating a Private Data Exchange

Currently, Data Exchange functionality is not enabled for all accounts. For you to follow the following instructions to set up a Private Data Exchange with data shares for another business unit or external partners, you must first contact Snowflake Support to have them enable the Data Exchange functionality. Also, remember that without performing replication, the Data Exchanges initially are set up only to share within a region.

Private Data Exchange Setup: Snowsight

Once you have worked with Snowflake Support and made sure that your Private Data Exchange functionality is enabled, then you will see additional views on the standard Data ➤ Shared Data view screen like Figure 14-10. Notice how there are two additional links for viewing including *Manage Exchanges* and *Reader Accounts*. Snowflake Support will have created something similar for your Private Data Exchange that will be in place of the IT Strategists Data Room listing.

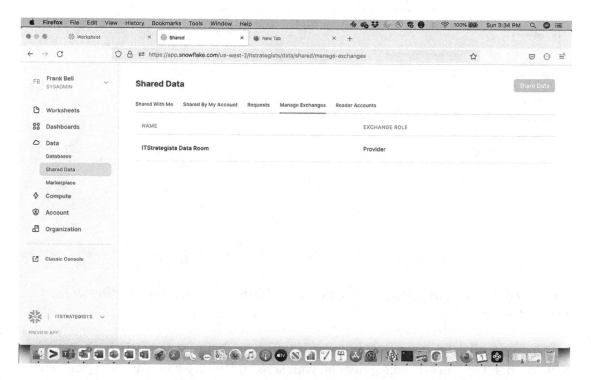

Figure 14-10. *Example of New Private Data Exchange – Snowsight*

Like the Snowflake Data Marketplace, you can now add both data provider accounts and data consumer accounts to your Private Data Exchange. If you take a step back, this is an enormous functionality that you can start to use to share data among all your ecosystem constituents from supplier partners to sales partners.

Previously, setting up the exchange had to be done mainly through commands, but now you can publish a data share as a data provider to your own Private Data Exchange or a partner's Private Data Exchange. Once you have access enabled, then instead of just the button *Share With Other Accounts*, you also will see an option Publish to <Your Private Data Exchange Name> as shown in Figure 14-11. Also, if you are enabled as a provider, you can create a provider profile as what you would do as a provider to the Snowflake Data Marketplace.

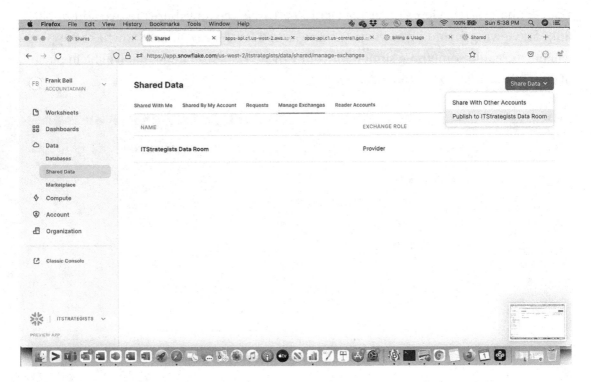

Figure 14-11. *Private Data Exchange – Publishing a Data Share Listing in Snowsight*

Once you click the equivalent button of Publish to <Your Private Data Exchange Name>, then Figure 14-12 will appear on your screen. This is where you can select what type of listing you want in the Private Data Exchange. Currently, there are two types of data shares including the free and instant access type [also named open data share listing] and the by request type [also named personalized data share listing].

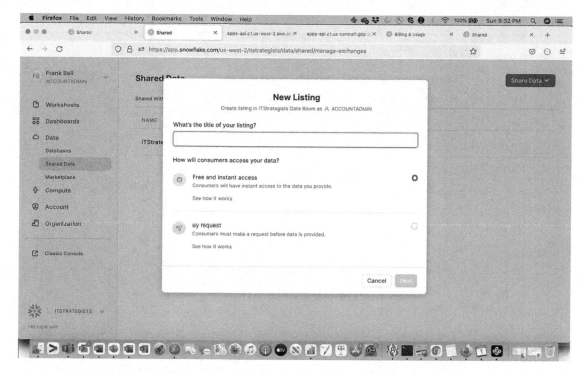

Figure 14-12. *Private Data Exchange – New Listing Options in Snowsight*

If you publish your provider listing as the by request type, then there is a short workflow involved to grant access to this type of provider listings. Again, it is the same as the Snowflake Data Marketplace listings workflow for the most part. You receive notifications of requests, and then you will need to go to an interface like Figure 14-13 where you can grant access to the listing to the specific account requesting it. Also, notice the *Requests* view has both your inbound and outbound requests, so if you acted as a data consumer and requested access to listings, you would see them there on this interface as you clicked the *Outbound* link.

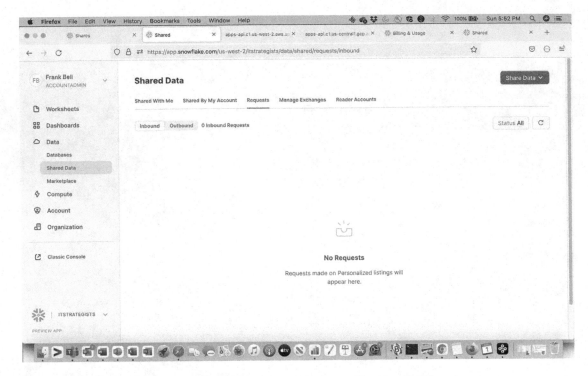

Figure 14-13. *Private Data Exchange – Requests Example View*

The Snowflake Data Marketplace

The Snowflake Data Marketplace is where Snowflake truly becomes the full Snowflake Data Cloud. At a high level, the Snowflake Data Marketplace is just a Private Data Exchange that is controlled by Snowflake as the administrator. Snowflake though is currently adding additional Data Marketplace functionality where they make this more of a platform. Before the Snowflake Data Marketplace came out in 2019, they had both Private Data Exchanges and an open data sharing exchange, which were revamped to their full Snowflake Data Marketplace in June 2020.

The Growth of the Snowflake Data Marketplace from June 2020 to June 2021

I have been tracking the Snowflake Data Marketplace since its official launch in June 2020. The growth over the 11 months from July 1, 2020 to June 1, 2021 was 371%. `https:// snowflakesolutions.net/snowflake-data-marketplace-growth-july-1-2020-to- june-1-2021/`. The Snowflake Data Marketplace continues to grow at a rapid pace both in providers and features. In 2021 Snowflake has added tools to directly buy marketplace data through the marketplace itself.

How to Use the Snowflake Data Marketplace as a Consumer

Snowflake Data Consumer Access and Navigation

Using the Snowflake Data Marketplace as a consumer has gotten much easier since June 2021. The initial Data Marketplace was revamped and improved and is even easier to access now via Snowsight. Let's jump right into the interface by going to this direct URL if you are already in the Snowsight interface: `https://app.snowflake.com/marketplace`. Or you can also access it from Snowsight by clicking Data ➤ Marketplace. Either way you should see a screen similar to Figure 14-14. Remember, this is Snowflake's Data Marketplace initial dashboard of listings, so it will vary based on what Snowflake decides to promote as the first screen view into the Snowflake Data Marketplace.

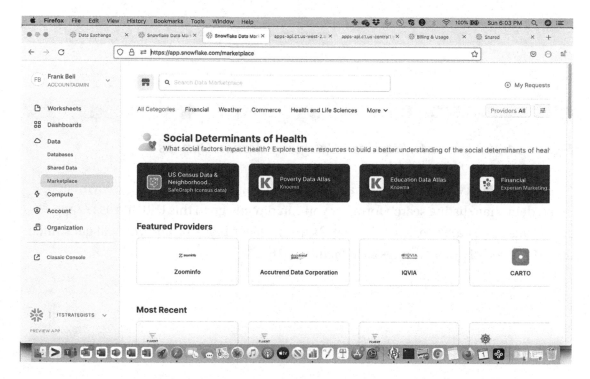

Figure 14-14. *Data Marketplace – Initial View*

You will notice in Figure 14-14 that you can navigate the hundreds of Snowflake data share listings by either selecting a category in the center or by selecting a specific data provider(s) to the right or using the filter selection to the right of the provider selection, and you can choose either from the *Ready to Query* or *Recently Published* filter. Figure 4-15 shows the category options available.

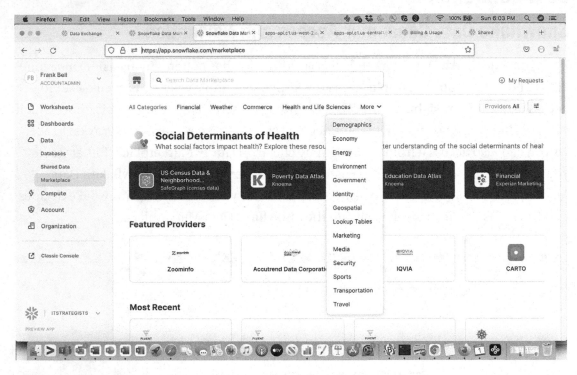

Figure 14-15. *Data Marketplace – Category Selection Options*

Let's see how we can search for a data share listing by entering "unemployment" in the data share listing search form. Or you can directly go to this URL `https://app.snowflake.com/marketplace/listings?search=unemployment`, and you will get similar data share search result listings as in Figure 14-16.

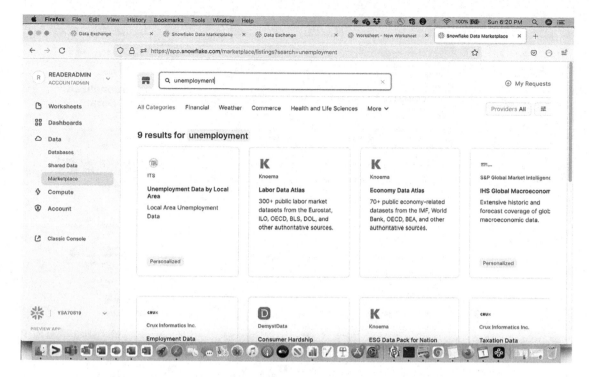

Figure 14-16. *Data Marketplace – How to Search Data Share Result Listings Example (Snowsight)*

Let's choose the first result with the red ITS logo and title of Unemployment Data by Local Area. This happens to be a data share that I maintain and will grant access to any *Snowflake Essentials* buyers. Figure 14-17 is what your interface should look like after you have selected the rectangular data share listing. You can also currently go directly to the URL `https://app.snowflake.com/marketplace/listing/GZSTZ62RSEH?search=une` `mployment.`

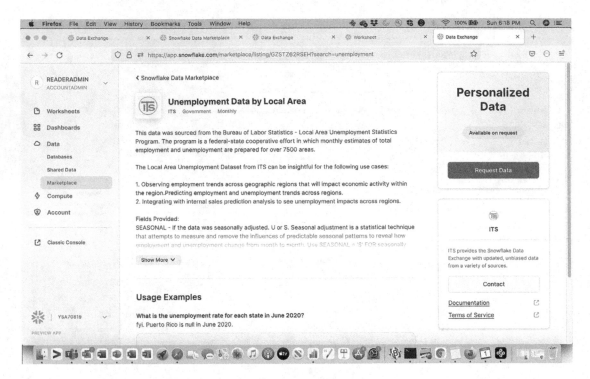

Figure 14-17. Data Marketplace – Individual "By Request" Unemployment by Local Area Data Share Listing (Snowsight)

If you read the previous section related to Private Data Exchange provider listings, then you will remember that there are currently two high-level types of data share listings [even though I understand Snowflake is working on some future types]. There is a by request type [also named personalized] and an open or free and instant access data share listing type. Until recently, they were either classified as personalized or open, even though some of the personalized data shares were not personalized but just needed to be requested. This recent change has made the marketplace easier to understand. This specific listing is a "by request" data share. Let's understand in detail how these two types of data shares differ.

Data Share Listings: By Request Type

Data shares that are listed as by request (also named personalized) require you to request them first, and then they go through a simple approval workflow for you to get access to them. Sometimes the data shares require some type of payment or fee to access. The Snowflake Data Marketplace team is currently working on some type of payment system, but for now the approval process for paid requests requires some type of human intervention and/or approval. For our *Snowflake Essentials* example though, let's just try it out and see how easy it is to request the Unemployment by Local Area by request–type listing. If you look at Figure 14-17, you just need to click that massive blue *Request Data* button to start the process. A new screen will pop up either with Figure 14-18 or a request for you to validate your email. You can only get to Figure 14-18 to enter the request details after your email has been validated. Once you get to a similar form as Figure 4-18, then just fill in the form fields of Company Name and Message (optional) fields and click the big blue *Request Data* button.

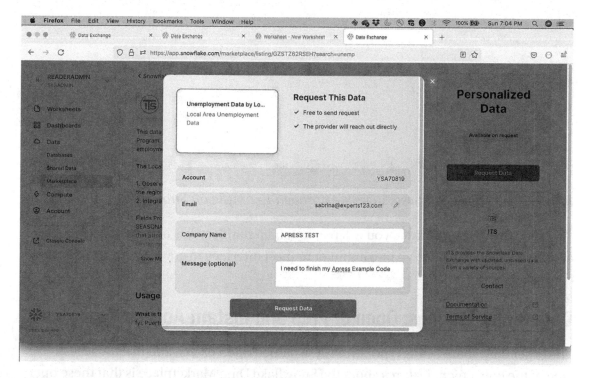

Figure 14-18. *Detailed Data Share Request Form – Snowsight*

As soon as you do that, the request gets put into the data share providers queue and view. You can also always check the current status of all of your data share requests by just going to Data ➤ Marketplace ➤ My Requests, or you can go directly in Snowsight to the URL `https://app.snowflake.com/marketplace/requests`.

Figure 14-19 shows an example of what my approved requests look like. If you only selected the one Unemployment by Local Area request, then you will just have one listing in the Pending view. Once you start requesting more data shares though, you can see the different requests by their status links of Approved, Denied, or Pending.

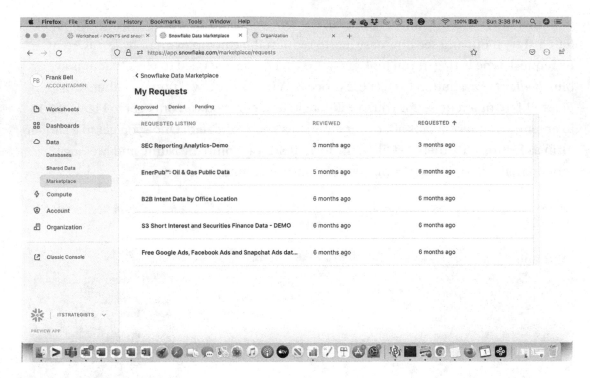

Figure 14-19. *My Requests View – Approved Example on Snowsight*

If you are a data provider, you will receive email notifications of data share requests, and then you will need to go into a similar request interface to grant access to your data consumers.

Data Share Listings: Open or Free and Instant Access Type

The other data share listing type currently is named open or free and instant access. One of the convenient features about the Snowflake Data Marketplace is that these open datasets are available for immediate access. Right now, from our tracking, it's about one-fifth of the marketplace is set up currently like this. What this means is you can instantly get access to these data shares very easily. Figure 4-20 shows an example of the Weather Source open and free dataset. You can also go directly to this dataset at the URL `https://app.snowflake.com/marketplace/listing/GZSNZ2LCLK?available=true`.

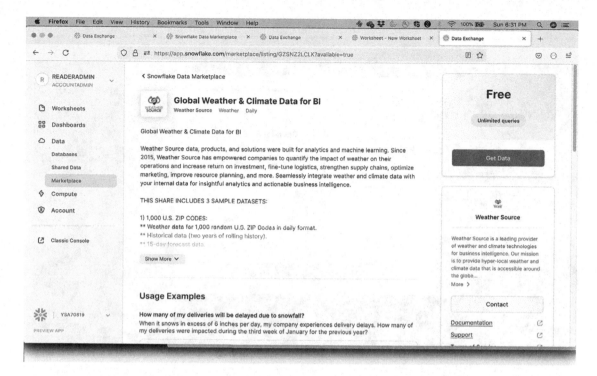

Figure 14-20. *Detailed Data Share "Get Data" Type – Global Weather – Snowsight*

Notice the major differences between the by request data share listing in Figure 4-17 and the free or open data share listing in Figure 14-20. The main difference is a big *Request Data* button vs. a *Get Data* button. So let's get access to our open data share here from Weather Source. All you need to do to start with is to click the big blue *Get Data* button. Then you will be prompted with an interface that looks like Figure 14-21 where you just need to decide on what you want to name your database based off this Weather Source data share. We always recommend coming up with a standard approach for naming databases tied to shares and matching the name to the share listing in some way. We also often use prefix standards such as S_ or SHARE_ to differentiate databases based on shares vs. standard databases.

Data Share Access

With that click of the *Get Data* buttons, you should get access if you meet all the criteria. At this point you can begin navigating to your databases based on the Weather Source data share or querying it immediately from your Snowsight worksheet.

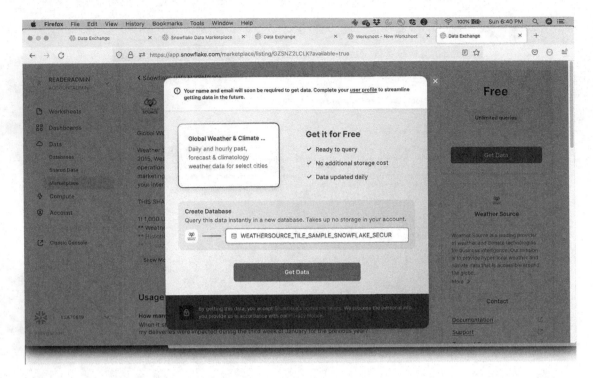

Figure 14-21. *Detailed Data Share Get Data Form (Snowsight)*

Summary

Snowflake data sharing is an amazing feature, which enables the Snowflake Data
Cloud to provide more than core cloud database functionality. The data sharing
functionality by itself is very powerful and differentiated from Snowflake's cloud
database competitors' offerings. Both the Private Data Exchange and Data Marketplace
functionality allow customers to take data and decision-making collaboration to
an entirely new level. Snowflake's data sharing with private exchanges and the
marketplace allows organizations to share data securely to almost an unlimited number
of constituents like we have never seen before. If utilized correctly by corporations
and organizations, they can transform their speed of business to make them more
competitive than ever before.

Index

A

Account management
  monitoring, 129
    ACCOUNTADMIN role, 130
    ad hoc query, 136, 137
    compute cluster costs, 131, 136
    cost, 129
    dashboard, 130
    database vendors, 129
    edition, 130
    fees, 130
    parameters, 132, 133
    primary areas, 130
    replication, 136
    resource monitor, 133, 135, 136
    storage costs, 131
    tools, 130
  risk mitigation, 118, 119
Amazon Web Services (AWS), 1, 17, 24
American National Standards Institute
    (ANSI), 197
Analytical cloud databases, 239
Antivirus, 141
Application programming interfaces
    (APIs), 300
AWS RedShift, 2

B

Big data architectural approaches
  data warehouse evolution, 27, 28
  hybrid architectures, 27

  shared nothing *vs.* shared disk, 27
Big data technology, 1

C

Caching architecture, 34, 35
Citi Bike NYC dataset, 253, 255
Classic Console
  account icon
    billing screen, 66
    policies, 67, 68
    Reader Accounts, 70–73
    resource monitors, 69, 70
    roles, 66, 67
    sessions, 68, 69
    usage screen, 65, 66
  databases, 40, 41
    file formats, 47–49
    pipes, 50–52
    schemas, 44, 45
    sequences, 50
    stages, 45–47
    tables, 42, 43
    views, 43, 44
  Data Marketplace, 56, 57
  data shares
    ACCOUNTADMIN role, 53
    creation, 55
    default view, 52, 53
    form entries, 55
    listings view, 53, 54
    Outbound link, 54
    parts, 54

© Frank Bell, Raj Chirumamilla, Bhaskar B. Joshi, Bjorn Lindstrom, Ruchi Soni, Sameer Videkar 2022
F. Bell et al., *Snowflake Essentials*, https://doi.org/10.1007/978-1-4842-7316-6

E

F

Q